BASICS OF BIOLOGICAL CHEMISTRY

BASICS OF BIOLOGICAL CHEMISTRY

Prabir Pujhari

ANMOL PUBLICATIONS PVT. LTD.
NEW DELHI - 110 002 (INDIA)

ANMOL PUBLICATIONS PVT. LTD.
H.O.: 4374/4B, Ansari Road, Darya Ganj,
New Delhi-110 002 (India)
Ph.: 23278000, 23261597
B.O.: No. 1015, Ist Main Road, BSK IIIrd Stage
IIIrd Phase, IIIrd Block,
Bangalore - 560 085 (India)
Visit us at: www.anmolpublications.com

Basics of Biological Chemistry

ISBN 978-81-261-3559-2

PRINTED IN INDIA

Printed at Mehra Offset Press, Delhi.

Contents

Preface

Essentials was written for students taking a broad introductory course in general, organic, and biological chemistry in a limited time. These students are often preparing for careers in allied health, such as nursing and various fields of therapy, but students in many other fields of study are often enrolled in this introductory chemistry course as well. This book assumes no background in chemistry, and students taking this course need only basic math skills.

This textbook has three primary objectives: (1) to present only that chemistry that is essential to the students' needs. The choice of material and depth of coverage was greatly influenced by the comments and suggestions of reviewers who have considerable experience in teaching general, organic, and biological chemistry; (2) to provide students with many examples of chemistry in health care and other applications to real life; and (3) to contribute to the broad educational experience of students by fostering analytical thinking and problem-solving skills.

Author

Chapter 1

The Molecular Design of Life

Part of a lipoprotein particle. A model of the structure of apolipoprotein A-I (yellow), shown surrounding sheets of lipids. The apolipoprotein is the major protein component of high-density lipoprotein particles in the blood. These particles are effective lipid transporters because the protein component provides an interface between the hydrophobic lipid chains and the aqueous environment of the bloodstream.

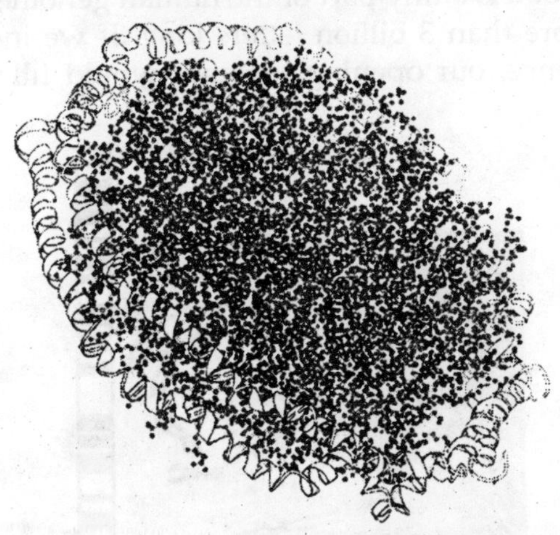

PRELUDE: BIOCHEMISTRY AND THE GENOMIC REVOLUTION

GACTTCACTTCTAATGATGATTATGGGAGAACTGGAGCCTT

CAGAGGGTAAAAATTAAGCACAGTGGAAGAATTTCATTC

TGTTCTCAGTTTTCCTGGATTATGCCTGGCACCATTAAAG

AAAATATCTTTGGTGTTTCCTATGATGAATATAGATACAG

AAGCGTCATCAAAGCATGCCAACTAGAAGAG....

This string of letters A, C, G, and T is a part of a DNA sequence. Since the biochemical techniques for DNA sequencing were first developed more than three decades ago, the genomes of dozens of organisms have been sequenced, and many more such sequences will be forthcoming. The information contained in these DNA sequences promises to shed light on many fascinating and important questions. What genes in *Vibrio cholera,* the bacterium that causes cholera, for example, distinguish it from its more benign relatives? How is the development of complex organisms controlled? What are the evolutionary relationships between organisms?

Sequencing studies have led us to a tremendous landmark in the history of biology and, indeed, humanity. *A nearly complete sequence of the entire human genome* has been determined. The string of As, Cs, Gs, and Ts with which we began this book is a tiny part of the human genome sequence, which is more than 3 billion letters long. If we included the entire sequence, our opening sentence would fill more than 500,000 pages.

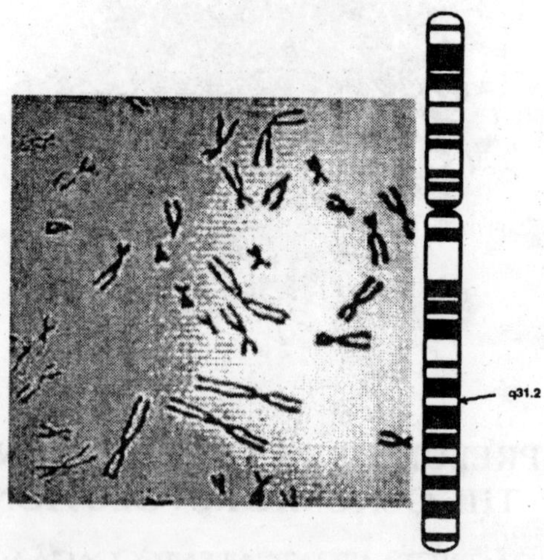

Fig. Disease and the genome

The implications of this knowledge cannot be overestimated. By using this blueprint for much of what it means to be human, scientists can begin the identification and characterization of sequences that foretell the appearance of specific diseases and particular physical attributes. One consequence will be the development of better means of diagnosing and treating diseases. Ultimately, physicians will be able to devise plans for preventing or managing heart disease or cancer that take account of individual variations. Although the sequencing of the human genome is an enormous step toward a complete understanding of living systems, much work needs to be done. Where are the functional genes within the sequence, and how do they interact with one another? How is the information in genes converted into the functional characteristics of an organism? Some of our goals in the study of biochemistry are to learn the concepts, tools, and facts that will allow us to address these questions. It is indeed an exciting time, the beginning of a new era in biochemistry.

DNA ILLUSTRATES THE RELATION BETWEEN FORM AND FUNCTION

The structure of DNA, an abbreviation for *deoxyribonucleicacid,* illustrates a basic principle common to all biomolecules: the intimate relation between structure and function. The remarkable properties of this chemical substance allow it to function as a very efficient and robust vehicle for storing information. We begin with an examination of the covalent structure of DNA and its extension into three dimensions.

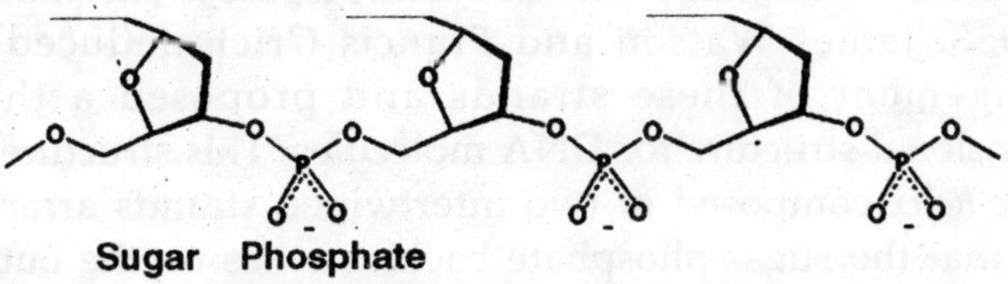

Sugar **Phosphate**

Fig. Covalent Structure of DNA

DNA IS CONSTRUCTED FROM FOUR BUILDING BLOCKS

DNA is a *linear polymer* made up of four different monomers. It has a fixed backbone from which protrude variable substituents. The backbone is built of repeating sugar-phosphate units. The sugars are molecules of *deoxyribose* from which DNA receives its name. Joined to each deoxyribose is one of four possible bases: adenine (A), cytosine (C), guanine (G), and thymine (T).

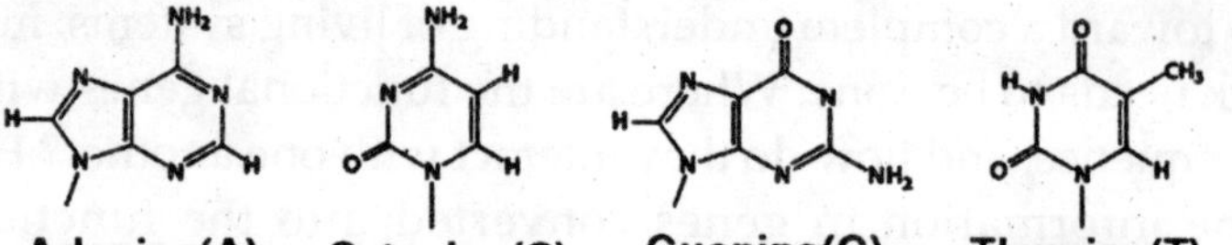

Adenine(A) Cytosine(C) Guanine(G) Thymine(T)

All four bases are planar but differ significantly in other respects. Thus, the monomers of DNA consist of a sugar-phosphate unit, with one of four bases attached to the sugar. *These bases can be arranged in any order along a strand of DNA.* The order of these bases is what is displayed in the sequence that begins this chapter. For example, the first base in the sequence shown is G (guanine), the second is A (adenine), and so on. *The sequence of bases along a DNA strand constitutes the genetic information*—the instructions for assembling proteins, which themselves orchestrate the synthesis of a host of other biomolecules that form cells and ultimately organisms.

TWO SINGLE STRANDS OF DNA COMBINE TO FORM A DOUBLE HELIX

Most DNA molecules consist of not one but two strands. How are these strands positioned with respect to one another? In 1953, James Watson and Francis Crick deduced the arrangement of these strands and proposed a three-dimensional structure for DNA molecules. This structure is a *double helix* composed of two intertwined strands arranged such that the sugar-phosphate backbone lies on the outside and the bases on the inside. The key to this structure is that the bases form *specific base pairs* (bp) held together by *hydrogen bonds*: adenine pairs with thymine (A-T) and guanine pairs with cytosine (G-C), as shown. Hydrogen bonds are much weaker than covalent bonds such as the carbon-carbon or carbon-nitrogen bonds that define the structures of the bases themselves. Such weak bonds are crucial to biochemical systems; they are weak enough to be reversibly broken in biochemical processes, yet they are strong enough, when many form simultaneously, to help stabilize specific structures such as the double helix.

The structure proposed by Watson and Crick has two properties of central importance to the role of DNA as the hereditary material. First, the structure is compatible with *any sequence of bases*. The base pairs have essentially the same shape and thus fit equally well into the centre of the double-helical structure. Second, because of base-pairing, *the sequence of bases along one strand completely determines the sequence along the other strand*.As Watson and Crick so coyly wrote: "It has not escaped our notice that the specific pairing we have postulated immediately suggests a possible copying mechanism for the genetic material." Thus, if the DNA double helix is separated into two single strands, each strand can act as a template for the generation of its partner strand through specific base-pair formation. *The three-dimensional structure of DNA beautifully illustrates the close connection between molecular form and function.*

RNA IS AN INTERMEDIATE IN THE FLOW OF GENETIC INFORMATION

An important nucleic acid in addition to DNA is *ribonucleicacid (RNA)*. Some viruses use RNA as the genetic material, and even those organisms that employ DNA must first convert the genetic information into RNA for the information to be accessible or functional. Structurally, RNA is quite similar to DNA. It is a linear polymer made up of a limited number of repeating monomers, each composed of a sugar, a phosphate, and a base. The sugar is ribose instead of deoxyribose (hence, *R*NA) and one of the bases is uracil (U) instead of thymine (T). Unlike DNA, an RNA molecule usually exists as a single strand, although significant segments within an RNA molecule may be double stranded, with G pairing primarily with C and A pairing with U. This intrastrand base-pairing generates RNA molecules with complex structures and activities, including catalysis.

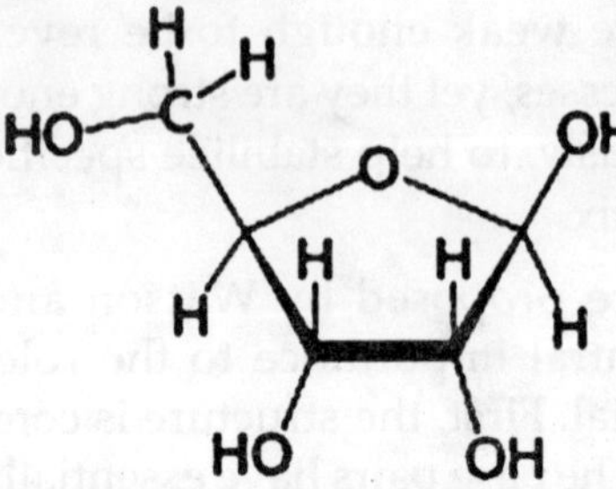

Ribose

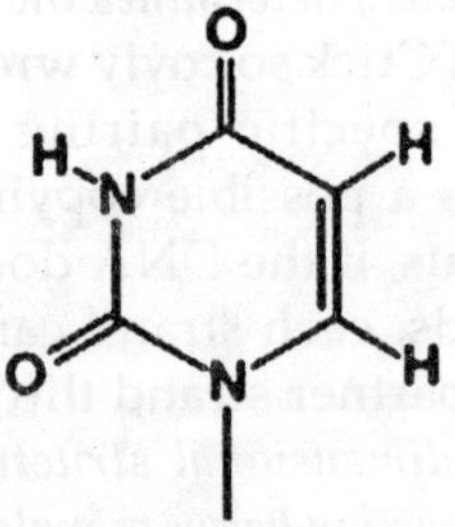

Uracil (U)

RNA has three basic roles in the cell. First, it serves as the intermediate in the flow of information from DNA to protein, the primary functional molecules of the cell. The DNA is copied, or *transcribed,* into messenger RNA (mRNA), and the mRNA is *translated* into protein. Second, RNA molecules serve as adaptors that translate the information in the nucleic acid sequence of mRNA into information designating the sequence of constituents that make up a protein. Finally, RNA molecules are important functional components of the molecular machinery, called ribosomes, that carries out the translation process. As will be discussed the unique position of RNA between the storage of genetic information in DNA and the functional expression of this information as protein as well as its potential to combine genetic and catalytic capabilities are indications that RNA played an important role in the evolution of life.

PROTEINS, ENCODED BY NUCLEIC ACIDS, PERFORM MOST CELL FUNCTIONS

A major role for many sequences of DNA is to encode the sequences of *proteins,* the workhorses within cells, participating in essentially all processes. Some proteins are key structural components, whereas others are specific catalysts (termed *enzymes*) that promote chemical reactions. Like DNA and RNA, proteins are linear polymers. However, proteins are more complicated in that they are formed from a selection of 20 building blocks, called *amino acids,* rather than 4.

The functional properties of proteins, like those of other biomolecules, are determined by their three-dimensional structures. Proteins possess an extremely important property: a protein spontaneously folds into a welldefined and elaborate three-dimensional structure that is dictated entirely by the sequence of amino acids along its chain. *The self-folding nature of proteins constitutes the transition from the one-dimensional world of sequence information to the three-dimensional world of biological function.* This marvelous ability of proteins to self assemble into complex structures is responsible for their dominant role in biochemistry.

How is the sequence of bases along DNA translated into a sequence of amino acids along a protein chain? We will consider the details of this process in later chapters, but the important finding is that *three bases along a DNA chain encode a single amino acid.* The specific correspondence between a set of three bases and 1 of the 20 amino acids is called the *genetic code.* Like the use of DNA as the genetic material, the genetic code is essentially universal; the same sequences of three bases encode the same amino acids in all life forms from simple microorganisms to complex, multicellular organisms such as human beings.

Knowledge of the functional and structural properties of proteins is absolutely essential to understanding the significance of the human genome sequence. For example, the sequence at the beginning of this chapter corresponds to a region of the genome that differs in people who have the genetic disorder *cystic fibrosis*. The most common mutation causing cystic fibrosis, the loss of three consecutive Ts from the gene sequence, leads to the loss of a single amino acid within a protein chain of 1480 amino acids. This seemingly slight difference—a loss of 1 amino acid of nearly 1500—creates a life-threatening condition. What is the normal function of the protein encoded by this gene? What properties of the encoded protein are compromised by this subtle defect? Can this knowledge be used to develop new treatments? These questions fall in the realm of biochemistry. Knowledge of the human genome sequence will greatly accelerate the pace at which connections are made between DNA sequences and disease as well as other human characteristics. However, these connections will be nearly meaningless without the knowledge of biochemistry necessary to interpret and exploit them.

BIOCHEMISTRY AND HUMAN BIOLOGY

Our understanding of biochemistry has had and will continue to have extensive effects on many aspects of human endeavor. *First, biochemistry is an intrinsically beautiful and fascinating body of knowledge.* We now know the essence and many of the details of the most fundamental processes in biochemistry, such as how a single molecule of DNA replicates to generate two identical copies of itself and how the sequence of bases in a DNA molecule determines the sequence of amino acids in an encoded protein. Our ability to describe these processes in detailed, mechanistic terms places a firm chemical foundation under other biological sciences. Moreover, the realization that we can understand essential life processes, such as the transmission of hereditary information, as chemical structures and their reactions has significant philosophical implications. What does it mean, biochemically, to be human? What are the biochemical differences between a human being, a chimpanzee, a mouse, and a fruit fly? Are we more similar than we are different?

Second, biochemistry is greatly influencing medicine and other fields. The molecular lesions causing sickle-cell anemia, cystic fibrosis, hemophilia, and many other genetic diseases have been elucidated at the biochemical level. Some of the molecular events that contribute to cancer development have been identified. An understanding of the underlying defects opens the door to the discovery of effective therapies. Biochemistry makes possible the rational design of new drugs, including specific inhibitors of enzymes required for the replication of viruses such as human immunodeficiency virus (HIV). Genetically engineered bacteria or other organisms can be used as "factories" to produce valuable proteins such as insulin and stimulators of blood-cell development. Biochemistry is also contributing richly to clinical diagnostics. For example, elevated levels of telltale enzymes in the blood reveal whether a patient has recently had a myocardial infarction (heart attack). DNA probes are coming into play in the precise diagnosis of inherited disorders, infectious diseases, and cancers. Agriculture, too, is benefiting from advances in biochemistry with the development of more effective, environmentally safer herbicides and pesticides and the creation of genetically engineered plants that are, for example, more resistant to insects. All of these endeavors are being accelerated by the advances in genomic sequencing.

Third, advances in biochemistry are enabling researchers to tackle some of the most exciting questions in biology and medicine. How does a fertilized egg give rise to cells as different as those in muscle, brain, and liver? How do the senses work? What are the molecular bases for mental disorders such as Alzheimer disease and schizophrenia? How does the immune system distinguish between self and nonself? What are the molecular mechanisms of short-term and long-term memory? The answers to such questions, which once seemed remote, have been partly uncovered and are likely to be more thoroughly revealed in the near future.

Because all living organisms on Earth are linked by a common origin, evolution provides a powerful organizing theme for biochemistry. This book is organized to emphasize the unifying principles revealed by evolutionary considerations. We begin in the next chapter with a brief tour along a plausible evolutionary path from the formation of some of the chemicals that we now associate with living organisms through the evolution of the processes essential for the development of complex, multicellular organisms.

The remainder of Part I of the book more fully introduces the most important classes of biochemicals as well as catalysis and regulation. Part II, Transducing and Storing Energy, describes how energy from chemicals or from sunlight is converted into usable forms and how this conversion is regulated. As we will see, a small set of molecules such as adenosine triphosphate (ATP) act as energy currencies that allow energy, however captured, to be utilized in a variety of biochemical processes. This part of the text examines the important pathways for the conversion of environmental energy into molecules such as ATP and uncovers many unifying principles. Part III, Synthesizing the Molecules of Life, illustrates the use of the molecules discussed in Part II to synthesize key molecular building blocks, such as the bases of DNA and amino acids, and then shows how these precursors are assembled into DNA, RNA, and proteins. In Parts II and III, we will highlight the relation between the reactions within each pathway and between those in different pathways so as to suggest how these individual reactions may have combined early in evolutionary history to produce the necessary molecules. From the student's perspective, the existence of features common to several pathways enables material mastered in one context to be readily applied to new contexts. Part IV, Responding to Environmental Changes, explores some of the mechanisms that cells and multicellular organisms have evolved to detect and respond to changes in the environment. The topics range from general mechanisms, common to all organisms, for regulating the expression of genes to the sensory systems used by human beings and other complex organisms. In many cases, we can now see how these elaborate systems evolved from pathways that existed earlier in evolutionary history. Many of the s in Part IV link biochemistry with other fields such as cell biology, immunology, and neuroscience. We are now ready to begin our journey into biochemistry with events that took place more than 3 billion years ago.

Chapter 2

Biochemical Evolution

Earth is approximately 4.5 billion years old. Remarkably, there is convincing fossil evidence that organisms morphologically (and very probably biochemically) resembling certain modern bacteria were in existence 3.5 billion years ago. With the use of the results of directed studies and accidental discoveries, it is now possible to construct a hypothetical yet plausible evolutionary path from the prebiotic world to the present. A number of uncertainties remain, particularly with regard to the earliest events. Nonetheless, a consideration of the steps along this path and the biochemical problems that had to be solved provides a useful perspective from which to regard the processes found in modern organisms. *These evolutionary connections make many aspects of biochemistry easier to understand.*

We can think of the path leading to modern living species as consisting of stages, although it is important to keep in mind that these stages were almost certainly not as distinct as presented here. The first stage was the initial generation of some of the key molecules of life—nucleic acids, proteins, carbohydrates, and lipids—by nonbiological processes. The second stage was fundamental—the transition from prebiotic chemistry to replicating systems. With the passage of time, these systems became increasingly sophisticated, enabling the formation of living cells. In the third stage, mechanisms evolved for interconverting energy from chemical sources and sunlight into forms that can be utilized to drive biochemical reactions. Intertwined with these energy-conversion processes are pathways for synthesizing the components of nucleic acids, proteins, and other key substances from simpler molecules. With the development of energy-conversion processes and biosynthetic pathways, a wide variety of unicellular organisms evolved. The fourth stage was the evolution of mechanisms that allowed cells to adjust their biochemistry to different, and often changing, environments. Organisms with these capabilities could form colonies comprising groups of interacting cells, and some eventually evolved into complex multicellular organisms.

This chapter introduces key challenges posed in the evolution of life, whose solutions are elaborated in later chapters. Exploring a possible evolutionary origin for these fundamental processes makes their use, in contrast with that of potential alternatives, more understandable.

KEY ORGANIC MOLECULES ARE USED BY LIVING SYSTEMS 3

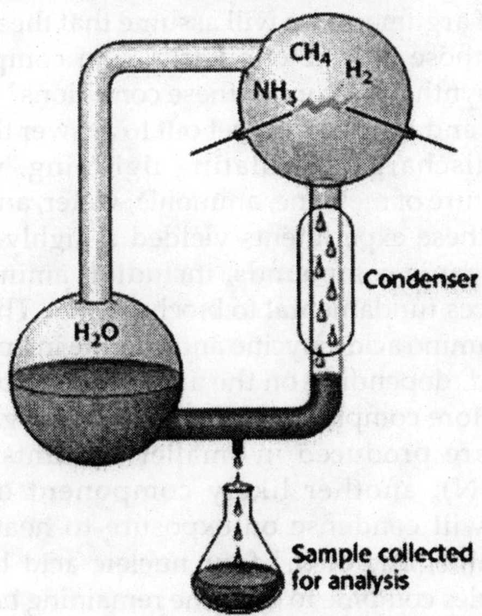

Fig. The Urey-Miller Experiment

Approximately 1 billion years after Earth's formation, life appeared, as already mentioned. Before life could exist, though, another major process needed to have taken place—the synthesis of the organic molecules required for living systems from simpler molecules found in the environment. The components of nucleic acids and proteins are relatively complex organic molecules, and one might expect that only sophisticated synthetic routes could produce them. However, this requirement appears not to have been the case. How did the building blocks of life come to be?

MANY COMPONENTS OF BIOCHEMICAL MACROMOLECULES CAN BE PRODUCED IN SIMPLE, PREBIOTIC REACTIONS

Among several competing theories about the conditions of the *prebiotic world*, none is completely satisfactory or problem-free. One theory holds that Earth's early atmosphere was highly reduced, rich in methane (CH_4), ammonia (NH_3), water (H_2O), and hydrogen (H_2), and that this atmosphere was subjected to large amounts of solar radiation and lightning. For the sake of argument, we will assume that these conditions were indeed those of prebiotic Earth. Can complex organic molecules be synthesized under these conditions? In the 1950s, Stanley Miller and Harold Urey set out to answer this question. An electric discharge, simulating lightning, was passed through a mixture of methane, ammonia, water, and hydrogen. Remarkably, these experiments yielded a highly nonrandom mixture of organic compounds, including amino acids and other substances fundamental to biochemistry. The procedure produces the amino acids glycine and alanine in approximately 2per cent yield, depending on the amount of carbon supplied as methane. More complex amino acids such as glutamic acid and leucine are produced in smaller amounts. Hydrogen cyanide (HCN), another likely component of the early atmosphere, will condense on exposure to heat or light to produce adenine, one of the four nucleic acid bases. Other simple molecules combine to form the remaining bases. A wide array of sugars, including ribose, can be formed from formaldehyde under prebiotic conditions.

UNCERTAINTIES OBSCURE THE ORIGINS OF SOME KEY BIOMOLECULES

The preceding observations suggest that many of the building blocks found in biology are unusually easy to synthesize and that significant amounts could have accumulated through the action of nonbiological processes. However, it is important to keep in mind that there are many uncertainties. For instance, ribose is just one of many sugars formed under prebiotic conditions. In addition, ribose is rather unstable under possible prebiotic conditions. Futhermore, ribose occurs in two mirror-image forms, only one of which occurs in modern RNA. To circumvent those problems, the first nucleic acid-like molecules have been suggested to have been bases attached to a different backbone and only later in evolutionary time was ribose incorporated to form nucleic acids as we know them today. Despite these uncertainties, an assortment of prebiotic molecules did arise in some fashion, and from this assortment *those with properties favourable for the processes that we now associate with life began to interact and to form more complicated compounds.*

EVOLUTION REQUIRES REPRODUCTION, VARIATION, AND SELECTIVE PRESSURE

Once the necessary building blocks were available, how did a living system arise and evolve? Before the appearance of life, simple molecular systems must have existed that subsequently evolved into the complex chemical systems that are characteristic of organisms. To address how this evolution occurred, we need to consider the *process* of evolution. There are several basic principles common to evolving systems, whether they are simple collections of molecules or competing populations of organisms. First, the most fundamental property of evolving systems is their ability to *replicate* or *reproduce.* Without this ability of *reproduction,* each "species" of molecule that might appear is doomed to extinction as soon as all its individual molecules degrade. For example, individual molecules of biological polymers such as ribonucleic acid are degraded by hydrolysis reactions and other processes. However, molecules that can replicate will continue to be represented in the population even if the lifetime of each individual molecule remains short.

A second principle fundamental to evolution is *variation*. The replicating systems must undergo changes. After all, if a system always replicates perfectly, the replicated molecule will always be the same as the parent molecule. Evolution cannot occur.

A third basic principle of evolution is *competition*. Replicating molecules compete with one another for available resources such as chemical precursors, and the competition allows the process of *evolution by natural selection* to occur. Variation will produce differing populations of molecules. Some variant offspring may, by chance, be better suited for survival and replication under the prevailing conditions than are their parent molecules. The prevailing conditions exert a *selective pressure* that gives an advantage to one of the variants. Those molecules that are best able to survive and to replicate themselves will increase in relative concentration. Thus, new molecules arise that are better able to replicate under the conditions of their environment. The same principles hold true for modern organisms. Organisms reproduce, show variation among individual organisms, and compete for resources; those variants with a selective advantage will reproduce more successfully. The changes leading to variation still take place at the molecular level, but the selective advantage is manifest at the organismal level.

THE PRINCIPLES OF EVOLUTION CAN BE DEMONSTRATED IN VITRO

Is there any evidence that evolution can take place at the molecular level? In 1967, Sol Spiegelman showed that replicating molecules could evolve new forms in an experiment that allowed him to observe molecular evolution in the test tube. He used as his evolving molecules RNA molecules derived from a bacterial virus called bacteriophage $Q\beta$. The genome of bacteriophage Q*β*, a single RNA strand of approximately 3300 bases, depends for its replication on the activity of a protein complex termed Q*β* replicase. Spiegelman mixed the replicase with a starting population of Q*β* RNA molecules. Under conditions in which there are ample amounts of precursors, no time constraints, and no other selective pressures, the composition of the population does not change from that of the parent molecules on replication. When selective pressures are applied, however, the composition of the population of molecules can change dramatically. For example, decreasing the time available for replication from 20 minutes to 5 minutes yielded, incrementally over 75 generations, a population of molecules dominated by a single species comprising only 550 bases. This species is replicated 15 times as rapidly as the parental Q*β* RNA. Spiegelman applied other selective pressures by, for example, limiting the concentrations of precursors or adding compounds that inhibit the replication process. In each case, new species appeared that replicated more effectively under the conditions imposed.

The process of evolution demonstrated in these studies depended on the existence of machinery for the replication of RNA fragments in the form of the Q*β* replicase. As noted earlier, one of the most elegant characteristics of nucleic acids is that the mechanism for their replication follows naturally from their molecular structure. This observation suggests that nucleic acids, perhaps RNA, could have become *self-replicating*. Indeed, the results of studies have revealed that single-stranded nucleic acids can serve as templates for the synthesis of their complementary strands and that this synthesis can occur spontaneously—that is, without biologically derived replication machinery. However, investigators have not yet found conditions in which an RNA molecule is fully capable of independent selfreplication from simple starting materials.

RNA MOLECULES CAN ACT AS CATALYSTS

The development of capabilities beyond simple replication required the generation of specific catalysts. A *catalyst* is a molecule that accelerates a particular chemical reaction without itself being chemically altered in the process. Some catalysts are highly specific; they promote certain reactions without substantially affecting closely related processes. Such catalysts allow the reactions of specific pathways to take place in preference to those of potential alternative pathways. Until the 1980s, all biological catalysts, termed *enzymes,* were believed to be proteins. Then, Tom Cech and Sidney Altman independently discovered that certain RNA molecules can be effective catalysts. These RNA catalysts have come to be known as *ribozymes.* The discovery of ribozymes suggested the possibility that catalytic RNA molecules could have played fundamental roles early in the evolution of life.

The catalytic ability of RNA molecules is related to their ability to adopt specific yet complex structures. This principle is illustrated by a "hammerhead" ribozyme, an RNA structure first identified in plant viruses. This RNA molecule promotes the cleavage of specific RNA molecules at specific sites; this cleavage is necessary for certain aspects of the viral life cycle. The ribozyme, which requires Mg^{2+} ion or other ions for the cleavage step to take place, forms a complex with its substrate RNA molecule that can adopt a reactive conformation. The existence of RNA molecules that possess specific binding and catalytic properties makes plausible the idea of an early *"RNA world"* inhabited by life forms dependent on RNA molecules to play all major roles, including those important in heredity, the storage of information, and the promotion of specific reactions—that is, biosynthesis and energy metabolism.

AMINO ACIDS AND THEIR POLYMERS CAN PLAY BIOSYNTHETIC AND CATALYTIC ROLES

In the early RNA world, the increasing populations of replicating RNA molecules would have consumed the building blocks of RNA that had been generated over long periods of time by prebiotic reactions. A shortage of these compounds would have favoured the evolution of alternative mechanisms for their synthesis. A large number of pathways are possible. Examining the biosynthetic routes utilized by modern organisms can be a source of insight into which pathways survived. A striking observation is that simple amino acids are used as building blocks for the RNA bases. For both purines (adenine and guanine) and pyrimidines (uracil and cytosine), an amino acid serves as a core onto which the remainder of the base is elaborated. In addition, nitrogen atoms are donated by the amino group of the amino acid aspartic acid and by the amide group of the glutamine side chain.

Amino acids are chemically more versatile than nucleic acids because their side chains carry a wider range of chemical functionality. Thus, amino acids or short polymers of amino acids linked by *peptide bonds,* called *polypeptides,* may have functioned as components of ribozymes to provide a specific reactivity. Furthermore, longer polypeptides are capable of spontaneously folding to form well-defined three-dimensional structures, dictated by the sequence of amino acids along their polypeptide chains. The ability of polypeptides to fold spontaneously into elaborate structures, which permit highly specific chemical interactions with other molecules, may have favoured the expansion of their roles in the course of evolution and is crucial to their dominant position in modern organisms. Today, most biological catalysts (enzymes) are not nucleic acids but are instead large polypeptides called *proteins*.

RNA TEMPLATE-DIRECTED POLYPEPTIDE SYNTHESIS LINKS THE RNA AND PROTEIN WORLDS

Polypeptides would have played only a limited role early in the evolution of life because their structures are not suited to self-replication in the way that nucleic acid structures are. However, polypeptides could have been included in evolutionary processes indirectly. For example, if the properties of a particular polypeptide favoured the survival and replication of a class of RNA molecules, then these RNA molecules could have evolved ribozyme activities that promoted the synthesis of that polypeptide. This method of producing polypeptides with specific amino acid sequences has several limitations. First, it seems likely that only relatively short specific polypeptides could have been produced in this manner. Second, it would have been difficult to accurately link the particular amino acids in the polypeptide in a reproducible manner. Finally, a different ribozyme would have been required for each polypeptide. A critical point in evolution was reached when an apparatus for polypeptide synthesis developed that allowed *the sequence of bases in an RNA molecule to directly dictate the sequence of amino acids in a polypeptide*. A code evolved that estabiished a relation between a specific sequence of three bases in RNA and an amino acid. We now call this set of three-base combinations, each encoding an amino acid, the *genetic code*. A decoding, or *translation*, system exists today as the *ribosome* and associated factors that are responsible for essentially all polypeptide synthesis from RNA templates in modern organisms.

An RNA molecule (*messenger RNA*, or *mRNA*), containing in its base sequence the information that specifies a particular protein, acts as a template to direct the synthesis of the polypeptide. Each amino acid is brought to the template attached to an adapter molecule specific to that amino acid. These adapters are specialized RNA molecules (called *transfer RNAs* or *tRNAs*). After initiation of the polypeptide chain, a tRNA molecule with its associated amino acid binds to the template through specific Watson-Crick base-pairing interactions. Two such molecules bind to the ribosome and peptide-bond formation is catalyzed by an RNA component (called *ribosomal RNA* or *rRNA*) of the ribosome. The first RNA departs (with neither the polypeptide chain nor an amino acid attached) and another tRNA with its associated amino acid bonds to the ribosome. The growing polypeptide chain is transferred to this newly bound amino acid with the formation of a new peptide bond. This cycle then repeats itself. This scheme allows the sequence of the RNA template to encode the sequence of the polypeptide and thereby makes possible the production of long polypeptides with specified sequences. Importantly, the ribosome is composed largely of RNA and is a highly sophisticated ribozyme, suggesting that it might be a surviving relic of the RNA world.

THE GENETIC CODE ELUCIDATES THE MECHANISMS OF EVOLUTION

The sequence of bases that encodes a functional protein molecule is called a *gene*. The genetic code—that is, the relation between the base sequence of a gene and the amino acid sequence of the polypeptide whose synthesis the gene directs—applies to all modern organisms with only very minor exceptions. This universality reveals that the genetic code was fixed early in the course of evolution and has been maintained to the present day.

We can now examine the mechanisms of evolution. Earlier, we considered how variation is required for evolution. We can now see that such variations in living systems are changes that alter the meaning of the genetic message. These variations are called *mutations*. A mutation can be as simple as a change in a single nucleotide (called a *point mutation*), such that a sequence of bases that encoded a particular amino acid may now encode another. A mutation can also be the insertion or deletion of several nucleotides.

Other types of alteration permit the more rapid evolution of new biochemical activities. For instance, entire s of the coding material can be duplicated, a process called *gene duplication*. One of the duplication products may accumulate mutations and eventually evolve into a gene with a different, but related, function. Furthermore, parts of a gene may be duplicated and added to parts of another to give rise to a completely new gene, which encodes a protein with properties associated with each parent gene. Higher organisms contain many large families of enzymes and other macromolecules that are clearly related to one another in the same manner. Thus, gene duplication followed by specialization has been a crucial process in evolution. It allows the generation of macromolecules having particular functions without the need to start from scratch. The accumulation of genes with subtle and large differences allows for the generation of more complex biochemical processes and pathways and thus more complex organisms.

TRANSFER RNAS ILLUSTRATE EVOLUTION BY GENE DUPLICATION

Transfer RNA molecules are the adaptors that associate an amino acid with its correct base sequence. Transfer RNA molecules are structurally similar to one another: each adopts a three-dimensional cloverleaf pattern of base-paired groups. Subtle differences in structure enable the protein-synthesis machinery to distinguish transfer RNA molecules with different amino acid specificities.

This family of related RNA molecules likely was generated by gene duplication followed by specialization. A nucleic acid sequence encoding one member of the family was duplicated, and the two copies evolved independently to generate molecules with specificities for different amino acids. This process was repeated, starting from one primordial transfer RNA gene until the 20 (or more) distinct members of the transfer RNA family present in modern organisms arose.

DNA IS A STABLE STORAGE FORM FOR GENETIC INFORMATION

It is plausible that RNA was utilized to store genetic information early in the history of life. However, in modern organisms (with the exception of some viruses), the RNA derivative DNA (*deoxy*ribonucleic acid) performs this function. The 2′ -hydroxyl group in the ribose unit of the RNA backbone is replaced by a hydrogen atom in DNA.

What is the selective advantage of DNA over RNA as the genetic material? The genetic material must be extremely stable so that sequence information can be passed on from generation to generation without degradation. RNA itself is a remarkably stable molecule; negative charges in the sugar-phosphate backbone protect it from attack by hydroxide ions that would lead to hydrolytic cleavage. However, the 2′ -hydroxyl group makes the RNA susceptible to base-catalyzed hydrolysis. The removal of the 2′ -hydroxyl group from the ribose decreases the rate of hydrolysis by approximately 100-fold under neutral conditions and perhaps even more under extreme conditions. Thus, the conversion of the genetic material from RNA into DNA would have substantially increased its chemical stability.

The evolutionary transition from RNA to DNA is recapitulated in the biosynthesis of DNA in modern organisms. In all cases, the building blocks used in the synthesis of DNA are synthesized from the corresponding building blocks of RNA by the action of enzymes termed *ribonucleotide reductases.* These enzymes convert ribonucleotides (a base and phosphate groups linked to a *ribose* sugar) into deoxyribonucleotides (a base and phosphates linked to *deoxyribose* sugar).

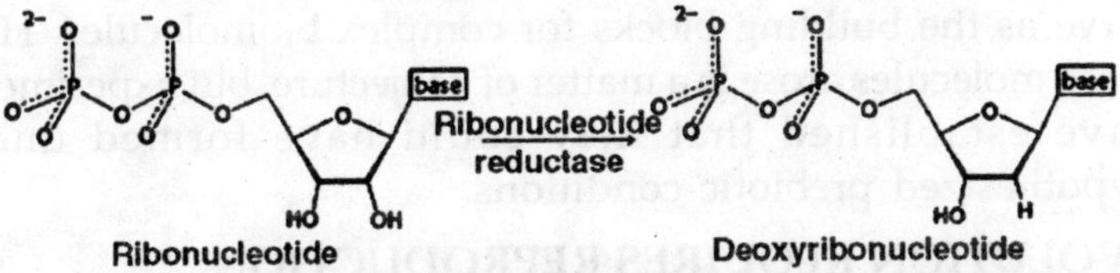

Ribonucleotide —Ribonucleotide reductase→ Deoxyribonucleotide

The properties of the ribonucleotide reductases vary substantially from species to species, but evidence suggests that they have a common mechanism of action and appear to have evolved from a common primordial enzyme.

The covalent structures of RNA and DNA differ in one other way. Whereas RNA contains *uracil,* DNA contains a methylated uracil derivative termed *thymine.* This modification also serves to protect the integrity of the genetic sequence, although it does so in a less direct manner. The methyl group present in thymine facilitates the repair of damaged DNA, providing an additional selective advantage.

Although DNA replaced RNA in the role of storing the genetic information, RNA maintained many of its other functions. RNA still provides the template that directs polypeptide synthesis, the adaptor molecules, the catalytic activity of the ribosomes, and other functions. Thus, the genetic message is *transcribed* from DNA into RNA and then *translated* into protein.

DNA $\xrightarrow{\text{Transcription}}$	RNA $\xrightarrow{\text{Translation}}$	polypeptide $\xrightarrow{\text{Folding}}$	Functional protein
Linear nucleic acid	Linear nucleic acid	Linear amino acid sequence	Three dimensional structure

This flow of sequence information from DNA to RNA to protein applies to all modern organisms.

KEY ORGANIC MOLECULES ARE USED BY LIVING SYSTEMS

The evolution of life required a series of transitions, beginning with the generation of organic molecules that could serve as the building blocks for complex biomolecules. How these molecules arose is a matter of conjecture, but experiments have established that they could have formed under hypothesized prebiotic conditions.

EVOLUTION REQUIRES REPRODUCTION, VARIATION, AND SELECTIVE PRESSURE

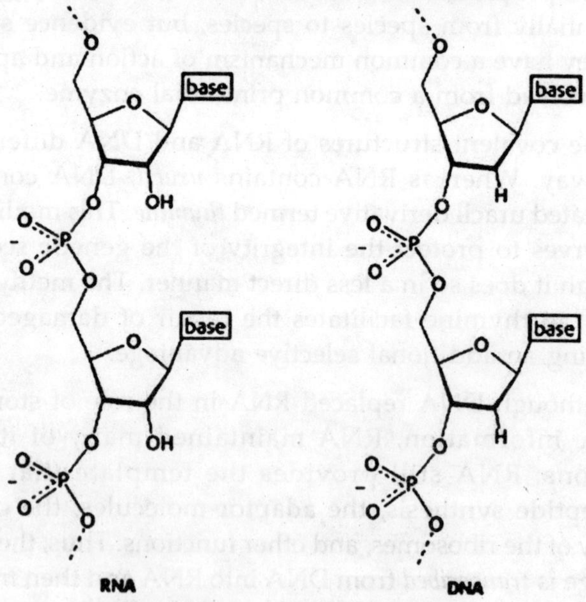

Fig. RNA and DNA Compared

The next major transition in the evolution of life was the formation of replicating molecules. Replication, coupled with variation and selective pressure, marked the beginning of evolution. Variation was introduced by a number of means, from simple base substitutions to the duplication of entire genes.

RNA appears to have been an early replicating molecule. Furthermore, some RNA molecules possess catalytic activity. However, the range of reactions that RNA is capable of catalyzing is limited. With time, the catalytic activity was transferred to proteins—linear polymers of the chemically versatile amino acids. RNA directed the synthesis of these proteins and still does in modern organisms through the development of a genetic code, which relates base sequence to amino acid sequence. Eventually, RNA lost its role as the gene to the chemically similar but more stable nucleic acid DNA. In modern organisms, RNA still serves as the link between DNA and protein.

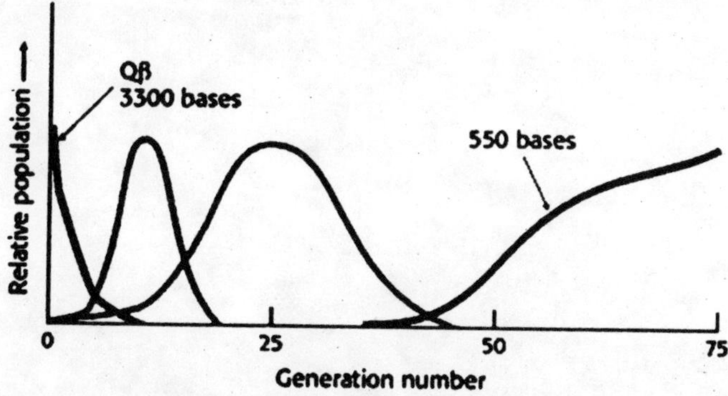

Fig. Evolution in a Test Tube

ENERGY TRANSFORMATIONS ARE NECESSARY TO SUSTAIN LIVING SYSTEMS

Another major transition in evolution was the ability to transform environmental energy into forms capable of being used by living systems. ATP serves as the cellular energy currency that links energy-yielding reactions with energy-requiring reactions. ATP itself is a product of the oxidation of fuel molecules, such as amino acids and sugars. With the evolution of membranes—hydrophobic barriers that delineate the borders of cells—ion gradients were required to prevent osmotic crises. These gradients were formed at the expense of ATP hydrolysis. Later, ion gradients generated by light or the oxidation of fuel molecules were used to synthesize ATP.

Chapter 3

Cells Can Respond to Changes in their Environments

The final transition was the evolution of sensing and signaling mechanisms that enabled a cell to respond to changes in its environment. These signaling mechanisms eventually led to cell-cell communication, which allowed the development of more-complex organisms. The record of much of what has occurred since the formation of primitive organisms is written in the genomes of extant organisms. Knowledge of these genomes and the mechanisms of evolution will enhance our understanding of the history of life on Earth as well as our understanding of existing organisms.

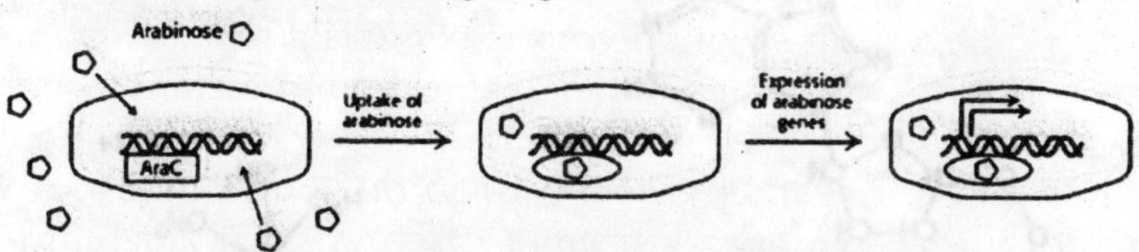

Fig. Responding to Environmental Conditions

CELLS CAN RESPOND TO CHANGES IN THEIR ENVIRONMENTS

The environments in which cells grow often change rapidly. For example, cells may consume all of a particular food source and must utilize others. To survive in a changing world, cells evolved mechanisms for adjusting their biochemistry in response to signals indicating environmental change. The adjustments can take many forms, including changes in the activities of preexisting enzyme molecules, changes in the rates of synthesis of new enzyme molecules, and changes in membrane-transport processes.

Initially, the detection of environmental signals occurred inside cells. Chemicals that could pass into cells, either by diffusion through the cell membrane or by the action of transport proteins, and could bind directly to proteins inside the cell and modulate their activities. An example is the use of the sugar arabinose by the bacterium *Escherichia coli*. *E. coli* cells are normally unable to use arabinose efficiently as a source of energy. However, if arabinose is their only source of carbon, *E. coli* cells synthesize enzymes that catalyze the conversion of this sugar into useful forms. This response is mediated by arabinose itself. If present in sufficient quantity outside the cell, arabinose can enter the cell through transport proteins. Once inside the cell, arabinose binds to a protein called AraC. This binding alters the structure of AraC so that it can now bind to specific sites in the bacterial DNA and increase RNA transcription from genes encoding enzymes that metabolize arabinose.

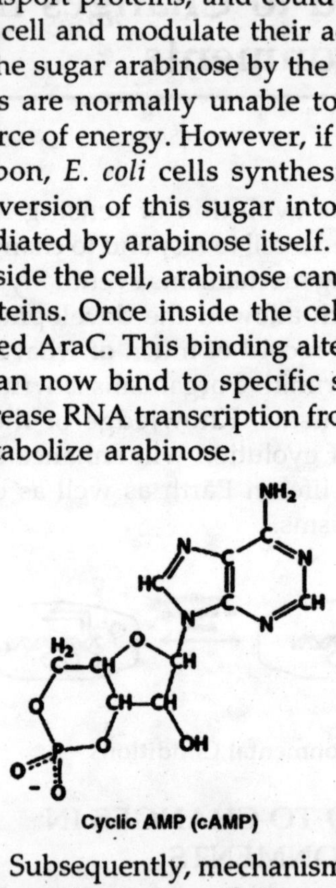

Cyclic AMP (cAMP)

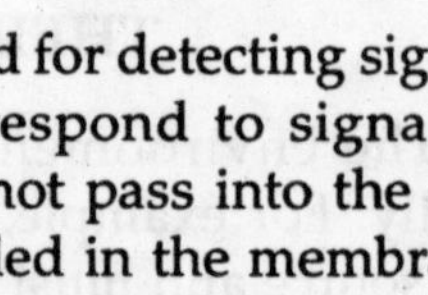

Calcium ion in water

Subsequently, mechanisms appeared for detecting signals at the cell surface. Cells could thus respond to signaling molecules even if those molecules did not pass into the cell. Receptor proteins evolved that, embedded in the membrane, could bind chemicals present in the cellular environment. Binding produced changes in the protein structure that could be detected at the inside surface of the cell membrane. By this means, chemicals outside the cell could influence events inside the cell. Many of these *signal-transduction pathways* make use of substances such as cyclic adenosine monophosphate (cAMP) and calcium ion as "second messengers" that can diffuse throughout the cell, spreading the word of environmental change.

The second messengers may bind to specific sensor proteins inside the cell and trigger responses such as the activation of enzymes.

FILAMENTOUS STRUCTURES AND MOLECULAR MOTORS ENABLE INTRACELLULAR AND CELLULAR MOVEMENT

The development of the ability to move was another important stage in the evolution of cells capable of adapting to a changing environment. Without this ability, nonphotosynthetic cells might have starved after consuming the nutrients available in their immediate vicinity.

Bacteria swim through the use of filamentous structures termed *flagella* that extend from their cell membranes. Each bacterial cell has several flagella, which, under appropriate conditions, form rotating bundles that efficiently propel the cell through the water. These flagella are long polymers consisting primarily of thousands of identical protein subunits. At the base of each flagellum are assemblies of proteins that act as motors to drive its rotation. The rotation of the flagellar motor is driven by the flow of protons from outside to inside the cell. Thus, energy stored in the form of a proton gradient is transduced into another form, rotatory motion.

Other mechanisms for motion, also depending on filamentous structures, evolved in other cells. The most important of these structures are *microfilaments* and *microtubules*. Microfilaments are polymers of the protein *actin*, and microtubules are polymers of two closely related proteins termed α- and *β-tubulin*. Unlike a bacterial flagellum, these filamentous structures are highly dynamic: they can rapidly increase or decrease in length through the addition or subtraction of component protein molecules. Microfilaments and microtubules also serve as tracks on which other proteins move, driven by the hydrolysis of ATP. Cells can change shape through the motion of *molecular motor proteins* along such filamentous structures that are changing in shape as a result of dynamic polymerization. Coordinated shape changes can be a means of moving a cell across a surface and are crucial to cell division. The motor proteins are also responsible for the transport of organelles and other structures within eukaryotic cells.

SOME CELLS CAN INTERACT TO FORM COLONIES WITH SPECIALIZED FUNCTIONS

Early organisms lived exclusively as single cells. Such organisms interacted with one another only indirectly by competing for resources in their environments. Certain of these organisms, however, developed the ability to form colonies comprising many interacting cells. In such groups, the environment of a cell is dominated by the presence of surrounding cells, which may be in direct contact with one another. These cells communicate with one another by a variety of signaling mechanisms and may respond to signals by altering enzyme activity or levels of gene expression. One result may be *cell differentiation;* differentiated cells are genetically identical but have different properties because their genes are expressed differently.

Several modern organisms are able to switch back and forth from existence as independent single cells to existence as multicellular colonies of differentiated cells. One of the most well characterized is the slime mold *Dictyostelium*. In favourable environments, this organism lives as individual cells; under conditions of starvation, however, the cells come together to form a cell aggregate. This aggregate, sometimes called a *slug,* can move as a unit to a potentially more favourable environment where it then forms a multicellular structure, termed a *fruiting body,* that rises substantially above the surface on which the cells are growing. Wind may carry cells released from the top of the fruiting body to sites where the food supply is more plentiful. On arriving in a well-stocked location, the cells grow, reproduce, and live as individual cells until the food supply is again exhausted.

The transition from unicellular to multicellular growth is triggered by cell-cell communication and reveals much about signaling processes between and within cells. Under starvation conditions, *Dictyostelium* cells release the signal molecule cyclic AMP. This molecule signals surrounding cells by binding to a membrane-bound protein receptor on the cell surface. The binding of cAMP molecules to these receptors triggers several responses, including movement in the direction of higher cAMP concentration, as well as the generation and release of additional cAMP molecules.

The cells aggregate by following cAMP gradients. Once in contact, they exchange additional signals and then differentiate into distinct *cell types,* each of which expresses the set of genes appropriate for its eventual role in forming the fruiting body. The life cycles of organisms such as *Dictyostelium* foreshadow the evolution of organisms that are multicellular throughout their lifetimes. It is also interesting to note the cAMP signals starvation in many organisms, including human beings.

THE DEVELOPMENT OF MULTICELLULAR ORGANISMS REQUIRES THE ORCHESTRATED DIFFERENTIATION OF CELLS

The fossil record indicates that macroscopic, multicellular organisms appeared approximately 600 million years ago. Most of the organisms familiar to us consist of many cells. For example, an adult human being contains approximately 100,000,000,000,000 cells. The cells that make up different organs are distinct and, even within one organ, many different cell types are present. Nonetheless, the DNA sequence in each cell is identical. The differences between cell types are the result of differences in how these genes are expressed.

Each multicellular organism begins as a single cell. For this cell to develop into a complex organism, the embryonic cells must follow an intricate programme of regulated gene expression, cell division, and cell movement. The developmental programme relies substantially on the responses of cells to the environment created by neighbouring cells. Cells in specific positions within the developing embryo divide to form particular tissues, such as muscle. Developmental pathways have been extensively studied in a number of organisms, including the nematode *Caenorhabditis elegans* a 1-mm-long worm containing 959 cells. A detailed map describing the fate of each cell in C. *elegans* from the fertilized egg to the adult is shown. Interestingly, proper development requires not only cell division but also the death of specific cells at particular points in time through a process called programmed cell death or *apoptosis*.

Investigations of genes and proteins that control development in a wide range of organisms have revealed a great many common features. Many of the molecules that control human development are evolutionarily related to those in relatively simple organisms such as C. *elegans*. Thus, solutions to the problem of controlling development in multicellular organisms arose early in evolution and have been adapted many times in the course of evolution, generating the great diversity of complex organisms.

THE UNITY OF BIOCHEMISTRY ALLOWS HUMAN BIOLOGY TO BE EFFECTIVELY PROBED THROUGH STUDIES OF OTHER ORGANISMS

All organisms on Earth have a common origin. How could complex organisms such as human beings have evolved from the simple organisms that existed at life's start? The path outlined in this chapter reveals that most of the fundamental processes of biochemistry were largely fixed early in the history of life. The complexity of organisms such as human beings is manifest, at a biochemical level, in the interactions between overlapping and competing pathways, which lead to the generation of intricately connected groups of specialized cells. The evolution of biochemical and physiological complexity is made possible by the effects of gene duplication followed by specialization. Paradoxically, the reliance on gene duplication also makes this complexity easier to comprehend. Consider, for example, the protein kinases—enzymes that transfer phosphoryl groups from ATP to specific amino acids in proteins. These enzymes play essential roles in many signal-transduction pathways and in the control of cell growth and differentiation. The human genome encodes approximately 500 proteins of this class; even a relatively simple, unicellular organism such as brewer's yeast has more than 100 protein kinases. Yet each of these enzymes is the evolutionary descendant of a common ancestral enzyme. Thus, *we can learn much about the essential behaviour of this large collection of proteins through studies of a single family member*. After the essential behaviour is understood, we can evaluate the specific adaptations that allow each family member to perform its particular biological functions.

Most central processes in biology have been characterized first in relatively simple organisms, often through a combination of genetic, physiological, and biochemical studies. Many of the processes controlling early embryonic development were elucidated by the results of studies of the fruit fly. The events controlling DNA replication and the cell cycle were first deciphered in yeast. Investigators can now test the functions of particular proteins in mammals by disrupting the genes that encode these proteins in mice and examining the effects. The investigations of organisms linked to us by common evolutionary pathways are powerful tools for exploring all of biology and for developing new understanding of normal human function and disease.

Chapter 4

Protein Structure and Function

Proteins are the most versatile macromolecules in living systems and serve crucial functions in essentially all biological processes. They function as catalysts, they transport and store other molecules such as oxygen, they provide mechanical support and immune protection, they generate movement, they transmit nerve impulses, and they control growth and differentiation. Indeed, much of this text will focus on understanding what proteins do and how they perform these functions.

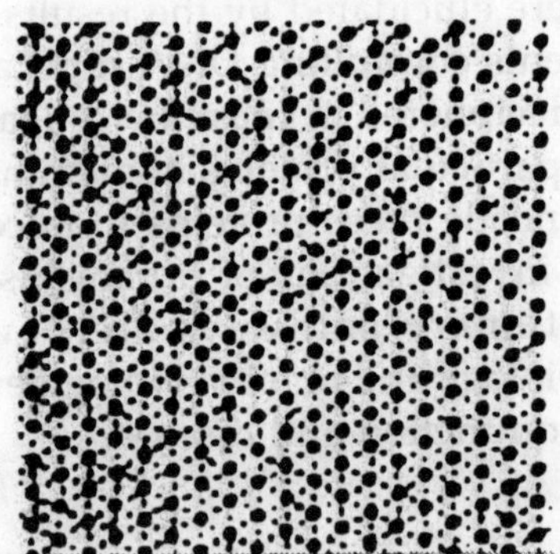

Fig. A Complex Protein Assembly

Several key properties enable proteins to participate in such a wide range of functions.

1. *Proteins are linear polymers built of monomer units called amino acids.* The construction of a vast array of macromolecules from a limited number of monomer

building blocks is a recurring theme in biochemistry. Does protein function depend on the linear sequence of amino acids? The function of a protein is directly dependent on its threedimensional structure. Remarkably, proteins spontaneously fold up into three-dimensional structures that are determined by the sequence of amino acids in the protein polymer. Thus, *proteins are the embodiment of the transition from the one-dimensional world of sequences to the three-dimensional world of molecules capable of diverse activities.*

2. *Proteins contain a wide range of functional groups.* These functional groups include alcohols, thiols, thioethers, carboxylic acids, carboxamides, and a variety of basic groups. When combined in various sequences, this array of functional groups accounts for the broad spectrum of protein function. For instance, the chemical reactivity associated with these groups is essential to the function of *enzymes*, the proteins that catalyze specific chemical reactions in biological systems.

3. *Proteins can interact with one another and with other biological macromolecules to form complex assemblies.* The proteins within these assemblies can act synergistically to generate capabilities not afforded by the individual component proteins. These assemblies include macro-molecular machines that carry out the accurate replication of DNA, the transmission of signals within cells, and many other essential processes.

4. *Some proteins are quite rigid, whereas others display limited flexibility.* Rigid units can function as structural elements in the cytoskeleton (the internal scaffolding within cells) or in connective tissue. Parts of proteins with limited flexibility may act as hinges, springs, and levers that are crucial to protein function, to the assembly of proteins with one another and with other molecules into complex units, and to the transmission of information within and between cells

PROTEINS ARE BUILT FROM A REPERTOIRE OF 20 AMINO ACIDS

Amino acids are the building blocks of proteins. An *α-amino acid* consists of a central carbon atom, called the *α carbon,* linked to an amino group, a carboxylic acid group, a hydrogen atom, and a distinctive R group. The R group is often referred to as the *side chain.* With four different groups connected to the tetrahedral α-carbon atom, α-amino acids are *chiral;* the two mirror-image forms are called the l isomer and the d isomer.

Only l amino acids are constituents of proteins. For almost all amino acids, the l isomer has *S* (rather than *R*) absolute configuration. Although considerable effort has gone into understanding why amino acids in proteins have this absolute configuration, no satisfactory explanation has been arrived at. It seems plausible that the selection of l over d was arbitrary but, once made, was fixed early in evolutionary history.

Amino acids in solution at neutral pH exist predominantly as *dipolar ions* (also called *zwitterions*). In the dipolar form, the amino group is protonated ($-NH_3^+$) and the carboxyl group is deprotonated ($-COO^-$). The ionization state of an amino acid varies with pH. In acid solution (e.g., pH 1), the amino group is protonated ($-NH_3^+$) and the carboxyl group is not dissociated (-COOH). As the pH is raised, the carboxylic acid is the first group to give up a proton, inasmuch as its pK_a is near 2. The dipolar form persists until the pH approaches 9, when the protonated amino group loses a proton.

Twenty kinds of side chains varying in *size, shape, charge, hydrogen-bonding capacity, hydrophobic character,* and *chemical reactivity* are commonly found in proteins. Indeed, all proteins in all species—bacterial, archaeal, and eukaryotic—are constructed from the same set of 20 amino acids. This fundamental alphabet of proteins is several billion years old. The remarkable range of functions mediated by proteins results from the diversity and versatility of these 20 building blocks.

Let us look at this set of amino acids. The simplest one is *glycine,* which has just a hydrogen atom as its side chain. With two hydrogen atoms bonded to the α-carbon atom, glycine is unique in being *achiral. Alanine,* the next simplest amino acid, has a methyl group (-CH_3) as its side chain.

Larger hydrocarbon side chains are found in *valine, leucine,* and *isoleucine. Methionine* contains a largely *aliphatic* side chain that includes a *thioether* (-S-) group. The side chain of isoleucine includes an additional chiral centre; only the isomer shown in figure is found in proteins. The larger aliphatic side chains are *hydrophobic*—that is, they tend to cluster together rather than contact water. The three-dimensional structures of water-soluble proteins are stabilized by this tendency of hydrophobic groups to come together, called *the hydrophobic effect.* The different sizes and shapes of these hydrocarbon side chains enable them to pack together to form compact structures with few holes. *Proline* also has an aliphatic side chain, but it differs from other members of the set of 20 in that its side chain is bonded to both the nitrogen and the α-carbon atoms. Proline markedly influences protein architecture because its ring structure makes it more conformationally restricted than the other amino acids.

Three amino acids with relatively simple *aromatic side chains* are part of the fundamental repertoire. *Phenylalanine,* as its name indicates, contains a phenyl ring attached in place of one of the hydrogens of alanine. The aromatic ring of *tyrosine* contains a hydroxyl group. This hydroxyl group is reactive, in contrast with the rather inert side chains of the other amino acids discussed thus far. *Tryptophan* has an indole ring joined to a methylene (-CH_2-) group; the indole group comprises two fused rings and an NH group. Phenylalanine is purely hydrophobic, whereas tyrosine and tryptophan are less so because of their hydroxyl and NH groups. The aromatic rings of tryptophan and tyrosine contain delocalized p electrons that strongly absorb ultraviolet light.

A compound's *extinction coefficient* indicates its ability to absorb light. Beer's law gives the absorbance (*A*) of light at a given wavelength:

$A = \varepsilon cl$ Beer's Law

where ε is the extinction coefficient [in units that are the reciprocals of molarity and distance in centimeters ($M^{-1}\ cm^{-1}$)], *c* is the concentration of the absorbing species (in units of molarity, M), and *l* is the length through which the light passes (in units of centimeters). For tryptophan, absorption is maximum at 280 nm and the extinction coefficient is 3400 $M^{-1}\ cm^{-1}$ whereas, for tyrosine, absorption is maximum at 276 nm and the extinction coefficient is a less-intense 1400 $M^{-1}\ cm^{-1}$. Phenylalanine absorbs light less strongly and at shorter wavelengths. The absorption of light at 280 nm can be used to estimate the concentration of a protein in solution if the number of tryptophan and tyrosine residues in the protein is known.

Two amino acids, *serine* and *threonine,* contain aliphatic *hydroxyl groups.* Serine can be thought of as a hydroxylated version of alanine, whereas threonine resembles valine with a hydroxyl group in place of one of the valine methyl groups. The hydroxyl groups on serine and threonine make them much more *hydrophilic* (water loving) and *reactive* than alanine and valine. Threonine, like isoleucine, contains an additional asymmetric centre; again only one isomer is present in proteins.

Cysteine is structurally similar to serine but contains a *sulfhydryl,* or *thiol* (-SH), group in place of the hydroxyl (-OH) group. The sulfhydryl group is much more reactive. Pairs of sulfhydryl groups may come together to form disulfide bonds, which are particularly important in stabilizing some proteins, as will be discussed shortly.

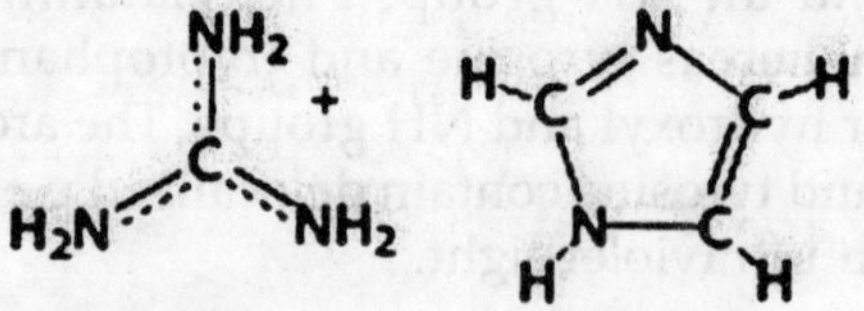

Guanidinium **Imidazole**

We turn now to amino acids with very polar side chains that render them highly hydrophilic. *Lysine* and *arginine* have relatively long side chains that terminate with groups that are *positively charged* at neutral pH. Lysine is capped by a primary amino group and arginine by a guanidinium group. *Histidine* contains an imidazole group, an aromatic ring that also can be positively charged.

With a pK_a value near 6, the imidazole group can be uncharged or positively charged near neutral pH, depending on its local environment. Indeed, histidine is often found in the active sites of enzymes, where the imidazole ring can bind and release protons in the course of enzymatic reactions.

The set of amino acids also contains two with *acidic side chains: aspartic acid* and *glutamic acid*. These amino acids are often called *aspartate* and *glutamate* to emphasize that their side chains are usually negatively charged at physiological pH. Nonetheless, in some proteins these side chains do accept protons, and this ability is often functionally important. In addition, the set includes uncharged derivatives of aspartate and glutamate—*asparagine* and *glutamine*—each of which contains a terminal *carboxamide* in place of a carboxylic acid.

Seven of the 20 amino acids have readily ionizable side chains. These 7 amino acids are able to donate or accept protons to facilitate reactions as well as to form ionic bonds.

Amino acids are often designated by either a three-letter abbreviation or a one-letter symbol. The abbreviations for amino acids are the first three letters of their names, except for asparagine (Asn), glutamine (Gln), isoleucine (Ile), and tryptophan (Trp). The symbols for many amino acids are the first letters of their names (e.g., G for glycine and L for leucine); the other symbols have been agreed on by convention. These abbreviations and symbols are an integral part of the vocabulary of biochemists.

How did this particular set of amino acids become the building blocks of proteins? First, as a set, they are diverse; their structural and chemical properties span a wide range, endowing proteins with the versatility to assume many functional roles. Second, as noted in figure many of these amino acids were probably available from prebiotic reactions. Finally, excessive intrinsic reactivity may have eliminated other possible amino acids. For example, amino acids such as homoserine and homocysteine tend to form five-membered cyclic forms that limit their use in proteins; the alternative amino acids that are found in proteins—serine and cysteine—do not readily cyclize, because the rings in their cyclic forms are too small.

Proteins are the workhorses of biochemistry, participating in essentially all cellular processes. Protein structure can be described at four levels. The primary structure refers to the amino acid sequence. The secondary structure refers to the conformation adopted by local regions of the polypeptide chain. Tertiary structure describes the overall folding of the polypeptide chain. Finally, quaternary structure refers to the specific association of multiple polypeptide chains to form multisubunit complexes.

PROTEINS ARE BUILT FROM A REPERTOIRE OF 20 AMINO ACIDS

Proteins are linear polymers of amino acids. Each amino acid consists of a central tetrahedral carbon atom linked to an amino group, a carboxylic acid group, a distinctive side chain, and a hydrogen. These tetrahedral centres, with the exception of that of glycine, are chiral; only the l isomer exists in natural proteins. All natural proteins are constructed from the same set of 20 amino acids. The side chains of these 20 building blocks vary tremendously in size, shape, and the presence of functional groups. They can be grouped as follows:

1. Aliphatic side chains—glycine, alanine, valine, leucine, isoleucine, methionine, and proline;
2. Aromatic side chains—phenylalanine, tyrosine, and tryptophan;
3. Hydroxyl-containing aliphatic side chains—serine and threonine;
4. Sulfhydryl-containing cysteine;

5. Basic side chains—lysine, arginine, and histidine;
6. Acidic side chains—aspartic acid and glutamic acid; and
7. Carboxamide-containing side chains—asparagine and glutamine. These groupings are somewhat arbitrary and many other sensible groupings are possible.

PRIMARY STRUCTURE: AMINO ACIDS ARE LINKED BY PEPTIDE BONDS TO FORM POLYPEPTIDE CHAINS

The amino acids in a polypeptide are linked by amide bonds formed between the carboxyl group of one amino acid and the amino group of the next. This linkage, called a peptide bond, has several important properties. First, it is resistant to hydrolysis so that proteins are remarkably stable kinetically. Second, the peptide group is planar because the C-N bond has considerable double-bond character. Third, each peptide bond has both a hydrogen-bond donor (the NH group) and a hydrogen-bond acceptor (the CO group). Hydrogen bonding between these backbone groups is a distinctive feature of protein structure. Finally, the peptide bond is uncharged, which allows proteins to form tightly packed globular structures having significant amounts of the backbone buried within the protein interior. Because they are linear polymers, proteins can be described as sequences of amino acids. Such sequences are written from the amino to the carboxyl terminus.

SECONDARY STRUCTURE: POLYPEPTIDE CHAINS CAN FOLD INTO REGULAR STRUCTURES SUCH AS THE ALPHA HELIX, THE BETA SHEET, AND TURNS AND LOOPS

Two major elements of secondary structure are the α helix and the β strand. In the *β* helix, the polypeptide chain twists into a tightly packed rod. Within the helix, the CO group of each amino acid is hydrogen bonded to the NH group of the amino acid four residues along the polypeptide chain. In the β strand, the polypeptide chain is nearly fully extended. Two or more β strands connected by NH-to-CO hydrogen bonds come together to form β sheets.

TERTIARY STRUCTURE: WATER-SOLUBLE PROTEINS FOLD INTO COMPACT STRUCTURES WITH NONPOLAR CORES

The compact, asymmetric structure that individual polypeptides attain is called tertiary structure. The tertiary structures of water-soluble proteins have features in common:

1. An interior formed of amino acids with hydrophobic side chains and
2. A surface formed largely of hydrophilic amino acids that interact with the aqueous environment. The driving force for the formation of the tertiary structure of water-soluble proteins is the hydrophobic interactions between the interior residues. Some proteins that exist in a hydrophobic environment, in membranes, display the inverse distribution of hydrophobic and hydrophilic amino acids. In these proteins, the hydrophobic amino acids are on the surface to interact with the environment, whereas the hydrophilic groups are shielded from the environment in the interior of the protein.

QUATERNARY STRUCTURE: POLYPEPTIDE CHAINS CAN ASSEMBLE INTO MULTISUBUNIT STRUCTURES

Proteins consisting of more than one polypeptide chain display quaternary structure, and each individual polypeptide chain is called a subunit. Quaternary structure can be as simple as two identical subunits or as complex as dozens of different subunits. In most cases, the subunits are held together by noncovalent bonds.

THE AMINO ACID SEQUENCE OF A PROTEIN DETERMINES ITS THREE-DIMENSIONAL STRUCTURE

The amino acid sequence completely determines the three-dimensional structure and, hence, all other properties of a protein. Some proteins can be unfolded completely yet refold efficiently when placed under conditions in which the folded form of the protein is stable. The amino acid sequence of a protein is determined by the sequences of bases in a DNA molecule. This one-dimensional sequence information is extended into the three-dimensional world by the ability of proteins to fold spontaneously. Protein folding is a highly cooperative process; structural intermediates between the unfolded and folded forms do not accumulate.

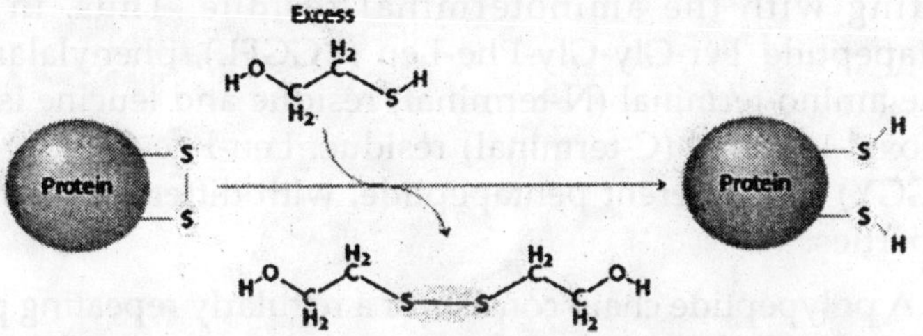

Excess

Fig. Role of β -Mercaptoethanol in Reducing Disulfide Bonds

The versatility of proteins is further enhanced by covalent modifications. Such modifications can incorporate functional groups not present in the 20 amino acids. Other modifications are important to the regulation of protein activity. Through their structural stability, diversity, and chemical reactivity, proteins make possible most of the key processes associated with life.

PRIMARY STRUCTURE: AMINO ACIDS ARE LINKED BY PEPTIDE BONDS TO FORM POLYPEPTIDE CHAINS

Proteins are *linear polymers* formed by linking the α-carboxyl group of one amino acid to the α-amino group of another amino acid with a *peptide bond* (also called an *amide bond*). The formation of a dipeptide from two amino acids is accompanied by the loss of a water molecule. The equilibrium of this reaction lies on the side of hydrolysis rather than synthesis. Hence, the biosynthesis of peptide bonds requires an input of free energy. Nonetheless, peptide bonds are quite *stable kinetically;* the lifetime of a peptide bond in aqueous solution in the absence of a catalyst approaches 1000 years.

A series of amino acids joined by peptide bonds form a *polypeptide chain,* and each amino acid unit in a polypeptide is called a *residue. A polypeptide chain has polarity* because its ends are different, with an α-amino group at one end and an α-carboxyl group at the other. By convention, *the amino end is taken to be the beginning of a polypeptide chain,* and so the sequence of amino acids in a polypeptide chain is written starting with the aminoterminal residue. Thus, in the pentapeptide Tyr-Gly-Gly-Phe-Leu (YGGFL), phenylalanine is the amino-terminal (N-terminal) residue and leucine is the carboxyl-terminal (C-terminal) residue. Leu-Phe-Gly-Gly-Tyr (LFGGY) is a different pentapeptide, with different chemical properties.

A polypeptide chain consists of a regularly repeating part, called the *main chain* or *backbone,* and a variable part, comprising the distinctive *side chains*. The polypeptide backbone is rich in hydrogen-bonding potential. Each residue contains a carbonyl group, which is a good hydrogen-bond acceptor and, with the exception of proline, an NH group, which is a good hydrogen-bond donor. These groups interact with each other and with functional groups from side chains to stabilize particular structures, as will be discussed in detail.

Most natural polypeptide chains contain between 50 and 2000 amino acid residues and are commonly referred to as *proteins*. Peptides made of small numbers of amino acids are called *oligopeptides* or simply *peptides*. The mean molecular weight of an amino acid residue is about 110, and so the molecular weights of most proteins are between 5500 and 220,000. We can also refer to the mass of a protein, which is expressed in units of daltons; one *dalton* is equal to one atomic mass unit. A protein with a molecular weight of 50,000 has a mass of 50,000 daltons, or 50 kd (kilodaltons).

In some proteins, the linear polypeptide chain is cross-linked. The most common cross-links are *disulfide bonds,* formed by the oxidation of a pair of cysteine residues. The resulting unit of linked cysteines is called *cystine*. Extracellular proteins often have several disulfide bonds, whereas intracellular proteins usually lack them. Rarely, nondisulfide cross-links derived from other side chains are present in some proteins. For example, collagen fibers in connective tissue are strengthened in this way, as are fibrin blood clots.

PROTEINS HAVE UNIQUE AMINO ACID SEQUENCES THAT ARE SPECIFIED BY GENES

In 1953, Frederick Sanger determined the amino acid sequence of insulin, a protein hormone. *This work is a landmark in biochemistry because it showed for the first time that a protein has a precisely defined amino acid sequence.* Moreover, it demonstrated that insulin consists only of l amino acids linked by peptide bonds between α-amino and α-carboxyl groups. This accomplishment stimulated other scientists to carry out sequence studies of a wide variety of proteins. Indeed, the complete amino acid sequences of more than 100,000 proteins are now known. *The striking fact is that each protein has a unique, precisely defined amino acid sequence.* The amino acid sequence of a protein is often referred to as its *primary structure.*

A series of incisive studies in the late 1950s and early 1960s revealed that the amino acid sequences of proteins are genetically determined. The sequence of nucleotides in DNA, the molecule of heredity, specifies a complementary sequence of nucleotides in RNA, which in turn specifies the amino acid sequence of a protein. In particular, each of the 20 amino acids of the repertoire is encoded by one or more specific sequences of three nucleotides.

Knowing amino acid sequences is important for several reasons. First, knowledge of the sequence of a protein is usually essential to elucidating its mechanism of action (e.g., the catalytic mechanism of an enzyme). Moreover, proteins with novel properties can be generated by varying the sequence of known proteins. Second, amino acid sequences determine the three-dimensional structures of proteins. Amino acid sequence is the link between the genetic message in DNA and the three-dimensional structure that performs a protein's biological function. Analyses of relations between amino acid sequences and three-dimensional structures of proteins are uncovering the rules that govern the folding of polypeptide chains. Third, sequence determination is a component of molecular pathology, a rapidly growing area of medicine. Alterations in amino acid sequence can produce abnormal function and disease. Severe and sometimes fatal diseases, such as sickle-cell anemia and cystic fibrosis, can result from a change in a single amino acid within a protein. Fourth, the sequence of a protein reveals much about its evolutionary history. Proteins resemble one another in amino acid sequence only if they have a common ancestor. Consequently, molecular events in evolution can be traced from amino acid sequences; molecular paleontology is a flourishing area of research.

POLYPEPTIDE CHAINS ARE FLEXIBLE YET CONFORMATIONALLY RESTRICTED

Examination of the geometry of the protein backbone reveals several important features. First, *the peptide bond is essentially planar*. Thus, for a pair of amino acids linked by a peptide bond, six atoms lie in the same plane: the α-carbon atom and CO group from the first amino acid and the NH group and α-carbon atom from the second amino acid. The nature of the chemical bonding within a peptide explains this geometric preference. The peptide bond has considerable *double-bond character*, which prevents rotation about this bond.

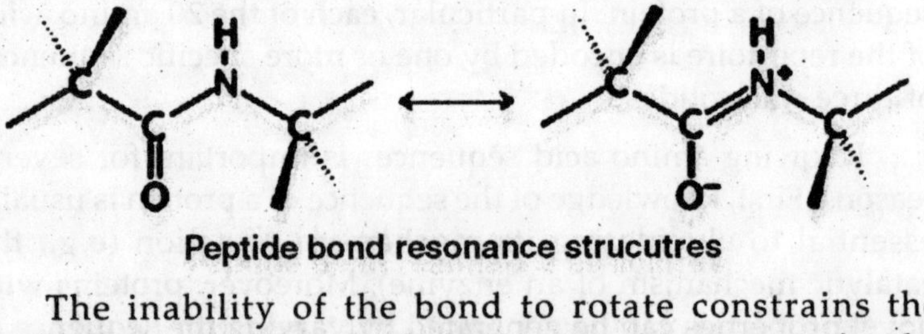

Peptide bond resonance strucutres

The inability of the bond to rotate constrains the conformation of the peptide backbone and accounts for the bond's planarity. This double-bond character is also expressed in the length of the bond between the CO and NH groups. The C-N distance in a peptide bond is typically 1.32 Å, which is between the values expected for a C-N single bond (1.49 Å) and a C=N double bond (1.27 Å), as shown in figure. Finally, the peptide bond is uncharged, allowing polymers of amino acids linked by peptide bonds to form tightly packed globular structures.

Two configurations are possible for a planar peptide bond. In the trans configuration, the two α-carbon atoms are on opposite sides of the peptide bond. In the cis configuration, these groups are on the same side of the peptide bond. *Almost all peptide bonds in proteins are trans.* This preference for trans over cis can be explained by the fact that steric clashes between groups attached to the α-carbon atoms hinder formation of the cis form but do not occur in the trans configuration. By far the most common cis peptide bonds are X-Pro linkages. Such bonds show less preference for the trans configuration because the nitrogen of proline is bonded to two tetrahedral carbon atoms, limiting the steric differences between the trans and cis forms.

In contrast with the peptide bond, the bonds between the amino group and the *α*-carbon atom and between the α-carbon atom and the carbonyl group are pure single bonds. The two adjacent rigid peptide units may rotate about these bonds, taking on various orientations. *This freedom of rotation about two bonds of each amino acid allows proteins to fold in many different ways.* The rotations about these bonds can be specified by dihedral angles. The angle of rotation about the bond between the nitrogen and the *α*-carbon atoms is cal!ed *phi* (φ). The angle of rotation about the bond between the *α*-carbon and the carbonyl carbon atoms is called *psi* (ψ). A clockwise rotation about either bond as viewed from the front of the back group corresponds to a positive value. The φ and y angles determine the path of the polypeptide chain.

Are all combinations of φ and ψ possible? G. N. Ramachandran recognized that many combinations are forbidden because of steric collisions between atoms. The allowed values can be visualized on a two-dimensional plot called a *Ramachandran diagram*. Three-quarters of the possible (φ, ψ) combinations are excluded simply by local steric clashes. *Steric exclusion, the fact that two atoms cannot be in the same place at the same time, can be a powerful organizing principle.*

The ability of biological polymers such as proteins to fold into welldefined structures is remarkable thermodynamically. Consider the equilibrium between an unfolded polymer that exists as a random coil—that is, as a mixture of many possible conformations—and the folded form that adopts a unique conformation. The favourable entropy associated with the large number of conformations in the unfolded form opposes folding and must be overcome by interactions favouring the folded form. Thus, highly flexible polymers with a large number of possible conformations do not fold into unique structures. The rigidity of the peptide unit and the restricted set of allowed φ and ψ angles limits the number of structures accessible to the unfolded form sufficiently to allow protein folding to occur.

SECONDARY STRUCTURE: POLYPEPTIDE CHAINS CAN FOLD INTO REGULAR STRUCTURES SUCH AS THE ALPHA HELIX, THE BETA SHEET, AND TURNS AND LOOPS

Can a polypeptide chain fold into a regularly repeating structure? In 1951, Linus Pauling and Robert Corey proposed two periodic structures called the *α helix* (*a*lpha helix) and the *β pleated sheet* (*beta* pleated sheet). Subsequently, other structures such as the *β turn* and *omega* (Ω) *loop* were identified. Although not periodic, these common turn or loop structures are well defined and contribute with α helices and β sheets to form the final protein structure.

THE ALPHA HELIX IS A COILED STRUCTURE STABILIZED BY INTRACHAIN HYDROGEN BONDS

In evaluating potential structures, Pauling and Corey considered which conformations of peptides were sterically allowed and which most fully exploited the hydrogen-bonding capacity of the backbone NH and CO groups. The first of their proposed structures, the *α helix*, is a rodlike structure. A tightly coiled backbone forms the inner part of the rod and the side chains extend outward in a helical array. The *α* helix is stabilized by hydrogen bonds between the NH and CO groups of the main chain. In particular, the CO group of each amino acid forms a hydrogen bond with the NH group of the amino acid that is situated four residues ahead in the sequence. Thus, except for amino acids near the ends of an α helix, all *the main-chain CO and NH groups are hydrogen bonded.* Each residue is related to the next one by a rise of 1.5 Å along the helix axis and a rotation of 100 degrees, which gives 3.6 amino acid residues per turn of helix. Thus, amino acids spaced three and four apart in the sequence are spatially quite close to one another in an α helix. In contrast, amino acids two apart in the sequence are situated on opposite sides of the helix and so are unlikely to make contact. The *pitch* of the α helix, which is equal to the product of the translation (1.5 Å) and the number of residues per turn (3.6), is 5.4 Å. The *screw sense* of a helix can be right-handed (clockwise) or left-handed (counterclockwise). The Ramachandran diagram reveals that both the right-handed and the left-handed helices are among allowed conformations. However, right-handed helices are energetically more favourable because there is less steric clash between the side chains and the backbone. *Essentially all α helices found in proteins are right-handed.* In schematic diagrams of proteins, α helices are depicted as twisted ribbons or rods.

Pauling and Corey predicted the structure of the α helix 6 years before it was actually seen in the x-ray reconstruction of the structure of myoglobin. *The elucidation of the structure of the α helix is a landmark in biochemistry because it demonstrated that the conformation of a polypeptide chain can be predicted if the properties of its components are rigorously and precisely known.*

The α-helical content of proteins ranges widely, from nearly none to almost 100per cent. For example, about 75per cent of the residues in ferritin, a protein that helps store iron, are in α helices. Single α helices are usually less than 45 Å long. However, two or more α helices can entwine to form a very stable structure, which can have a length of 1000 Å (100 µm, or 0.1µm) or more. Such *α-helical coiled coils* are found in myosin and tropomyosin in muscle, in fibrin in blood clots, and in keratin in hair. The helical cables in these proteins serve a mechanical role in forming stiff bundles of fibers, as in porcupine quills. The cytoskeleton (internal scaffolding) of cells is rich in so-called intermediate filaments, which also are two-stranded α-helical coiled coils. Many proteins that span biological membranes also contain α helices.

BETA SHEETS ARE STABILIZED BY HYDROGEN BONDING BETWEEN POLYPEPTIDE STRANDS

Pauling and Corey discovered another periodic structural motif, which they named the *β pleated sheet* (β because it was the second structure that they elucidated, the *α* helix having been the first). The β pleated sheet (or, more simply, the β sheet) differs markedly from the rodlike *α* helix. A polypeptide chain, called a *β strand,* in a β sheet is almost fully extended rather than being tightly coiled as in the α helix. A range of extended structures are sterically allowed.

The distance between adjacent amino acids along a β strand is approximately 3.5 Å, in contrast with a distance of 1.5 Å along an α helix. The side chains of adjacent amino acids point in opposite directions. A β sheet is formed by linking two or more β strands by hydrogen bonds. Adjacent chains in a β sheet can run in opposite directions (antiparallel β sheet) or in the same direction (parallel β sheet). In the antiparallel arrangement, the NH group and the CO group of each amino acid are respectively hydrogen bonded to the CO group and the NH group of a partner on the adjacent chain. In the parallel arrangement, the hydrogen-bonding scheme is slightly more complicated. For each amino acid, the NH group is hydrogen bonded to the CO group of one amino acid on the adjacent strand, whereas the CO group is hydrogen bonded to the NH group on the amino acid two residues farther along the chain. Many strands, typically 4 or 5 but as many as 10 or more, can come together in β sheets. Such β sheets can be purely antiparallel, purely parallel, or mixed.

In schematic diagrams, β strands are usually depicted by broad arrows pointing in the direction of the carboxyl-terminal end to indicate the type of β sheet formed—parallel or antiparallel. More structurally diverse than α helices, β sheets can be relatively flat but most adopt a somewhat twisted shape. The β sheet is an important structural element in many proteins. For example, fatty acid-binding proteins, important for lipid metabolism, are built almost entirely from β sheets.

POLYPEPTIDE CHAINS CAN CHANGE DIRECTION BY MAKING REVERSE TURNS AND LOOPS

Most proteins have compact, globular shapes, requiring reversals in the direction of their polypeptide chains. Many of these reversals are accomplished by a common structural element called the *reverse turn* (also known as the *β turn* or *hairpin bend*), illustrated in figure. In many reverse turns, the CO group of residue *i* of a polypeptide is hydrogen bonded to the NH group of residue *i* + 3. This interaction stabilizes abrupt changes in direction of the polypeptide chain. In other cases, more elaborate structures are responsible for chain reversals. These structures are called *loops* or sometimes *Ω loops* (omega loops) to suggest their overall shape. Unlike α helices and β strands, loops do not have regular, periodic structures. Nonetheless, loop structures are often rigid and well defined. Turns and loops invariably lie on the surfaces of proteins and thus often participate in interactions between proteins and other molecules. The distribution of α helices, β strands, and turns along a protein chain is often referred to as its *secondary structure.*

TERTIARY STRUCTURE: WATER-SOLUBLE PROTEINS FOLD INTO COMPACT STRUCTURES WITH NONPOLAR CORES

Let us now examine how amino acids are grouped together in a complete protein. X-ray crystallographic and nuclear magnetic resonance studies have revealed the detailed three-dimensional structures of thousands of proteins. We begin here with a preview of *myoglobin,* the first protein to be seen in atomic detail.

Myoglobin, the oxygen carrier in muscle, is a single polypeptide chain of 153 amino acids. The capacity of myoglobin to bind oxygen depends on the presence of *heme,* a nonpolypeptide *prosthetic (helper) group* consisting of protoporphyrin IX and a central iron atom. *Myo-globin is an extremely compact molecule.* Its overall dimensions are 45 × 35 × 25 Å, an order of magnitude less than if it were fully stretched out. About 70per cent of the main chain is folded into eight α helices, and much of the rest of the chain forms turns and loops between helices.

The folding of the main chain of myoglobin, like that of most other proteins, is complex and devoid of symmetry. The overall course of the polypeptide chain of a protein is referred to as its *tertiary structure.* A unifying principle emerges from the distribution of side chains. The striking fact is that *the interior consists almost entirely of nonpolar residues* such as leucine, valine, methionine, and phenylalanine. Charged residues such as aspartate, glutamate, lysine, and arginine are absent from the inside of myoglobin. The only polar residues inside are two histidine residues, which play critical roles in binding iron and oxygen. The outside of myoglobin, on the other hand, consists of both polar and nonpolar residues. The spacefilling model shows that there is very little empty space inside.

This contrasting distribution of polar and nonpolar residues reveals a key facet of protein architecture. In an aqueous environment, protein folding is driven by the strong tendency of hydrophobic residues to be excluded from water. Recall that a system is more thermodynamically stable when hydrophobic groups are clustered rather than extended into the aqueous surroundings. *The polypeptide chain therefore folds so that its hydrophobic side chains are buried and its polar, charged chains are on the surface.* Many α helices and β strands are amphipathic; that is, the α helix or β strand has a hydrophobic face, which points into the protein interior, and a more polar face, which points into solution. The fate of the main chain accompanying the hydrophobic side chains is important, too. An unpaired peptide NH or CO group markedly prefers water to a nonpolar milieu. The secret of burying a segment of main chain in a hydrophobic environment is pairing all the NH and CO groups by hydrogen bonding. This pairing is neatly accomplished in an α helix or β sheet. Van der Waals interactions between tightly packed hydrocarbon side chains also contribute to the stability of proteins. We can now understand why the set of 20 amino acids contains several that differ subtly in size and shape. They provide a palette from which to choose to fill the interior of a protein neatly and thereby maximize van der Waals interactions, which require intimate contact.

Some proteins that span biological membranes are "the exceptions that prove the rule" regarding the distribution of hydrophobic and hydrophilic amino acids throughout three-dimensional structures. For example, consider porins, proteins found in the outer membranes of many bacteria. The permeability barriers of membranes are built largely of alkane chains that are quite hydrophobic. Thus, porins are covered on the outside largely with hydrophobic residues that interact with the neighbouring alkane chains. In contrast, the centre of the protein contains many charged and polar amino acids that surround a water-filled channel running through the middle of the protein. Thus, because porins function in hydrophobic environments, they are "inside out" relative to proteins that function in aqueous solution.

Some polypeptide chains fold into two or more compact regions that may be connected by a flexible segment of polypeptide chain, rather like pearls on a string. These compact globular units, called *domains,* range in size from about 30 to 400 amino acid residues. For example, the extracellular part of CD4, the cell-surface protein on certain cells of the immune system to which the human immunodeficiency virus (HIV) attaches itself, comprises four similar domains of approximately 100 amino acids each. Often, proteins are found to have domains in common even if their overall tertiary structures are different.

QUATERNARY STRUCTURE: POLYPEPTIDE CHAINS CAN ASSEMBLE INTO MULTISUBUNIT STRUCTURES

Four levels of structure are frequently cited in discussions of protein architecture. So far, we have considered three of them. *Primary structure* is the amino acid sequence. *Secondary structure* refers to the spatial arrangement of amino acid residues that are nearby in the sequence. Some of these arrangements are of a regular kind, giving rise to a periodic structure. The α helix and β strand are elements of secondary structure. *Tertiary structure* refers to the spatial arrangement of amino acid residues that are far apart in the sequence and to the pattern of disulfide bonds. We now turn to proteins containing more than one polypeptide chain. Such proteins exhibit a fourth level of structural organization. Each polypeptide chain in such a protein is called a *subunit*. *Quaternary structure* refers to the spatial arrangement of subunits and the nature of their interactions. The simplest sort of quaternary structure is a *dimer*, consisting of two identical subunits. This organization is present in the DNA-binding protein Cro found in a bacterial virus called λ. More complicated quaternary structures also are common. More than one type of subunit can be present, often in variable numbers. For example, human hemoglobin, the oxygen-carrying protein in blood, consists of two subunits of one type (designated α) and two subunits of another type (designated β). Thus, the hemoglobin molecule exists as an $\alpha_2\beta_2$ tetramer. Subtle changes in the arrangement of subunits within the hemoglobin molecule allow it to carry oxygen from the lungs to tissues with great efficiency.

Viruses make the most of a limited amount of genetic information by forming coats that use the same kind of subunit repetitively in a symmetric array. The coat of rhinovirus, the virus that causes the common cold, includes 60 copies each of four subunits. The subunits come together to form a nearly spherical shell that encloses the viral genome.

THE AMINO ACID SEQUENCE OF A PROTEIN DETERMINES ITS THREE-DIMENSIONAL STRUCTURE

How is the elaborate three-dimensional structure of proteins attained, and how is the three-dimensional structure related to the one-dimensional amino acid sequence information? The classic work of Christian Anfinsen in the 1950s on the enzyme ribonuclease revealed the relation between the amino acid sequence of a protein and its conformation. Ribonuclease is a single polypeptide chain consisting of 124 amino acid residues cross-linked by four disulfide bonds. Anfinsen's plan was to destroy the three-dimensional structure of the enzyme and to then determine what conditions were required to restore the structure.

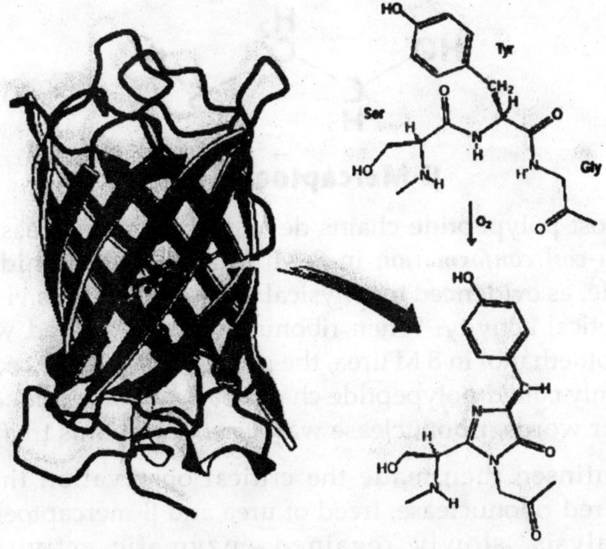

Fig. Chemical Rearrangement in GFP

Agents such as urea or guanidinium chloride effectively disrupt the noncovalent bonds, although the mechanism of action of these agents is not fully understood. The disulfide bonds can be cleaved reversibly by reducing them with a reagent such as *β-mercaptoethanol*. In the presence of a large excess of β-mercaptoethanol, a protein is produced in which the disulfides (cystines) are fully converted into sulfhydryls (cysteines).

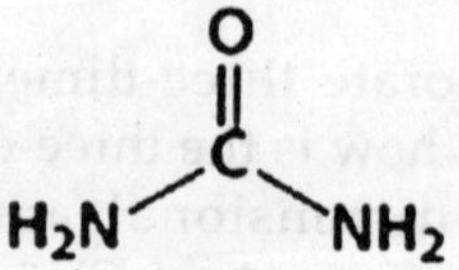

Urea

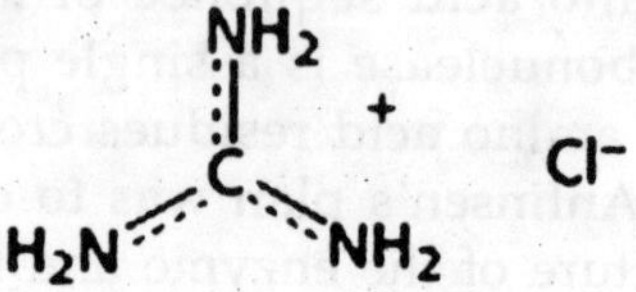

Guanidinium chloride

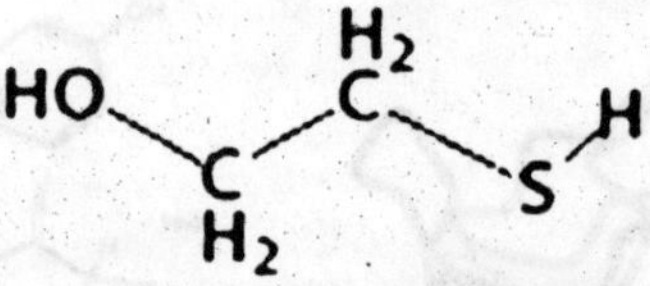

β-Mercaptoethanol

Most polypeptide chains devoid of cross-links assume a *random-coil conformation* in 8 M urea or 6 M guanidinium chloride, as evidenced by physical properties such as viscosity and optical activity. When ribonuclease was treated with β-mercaptoethanol in 8 M urea, the product was a fully reduced, randomly coiled polypeptide chain *devoid of enzymatic activity*. In other words, ribonuclease was *denatured* by this treatment.

Anfinsen then made the critical observation that the denatured ribonuclease, freed of urea and β-mercaptoethanol by dialysis, slowly regained enzymatic activity. He immediately perceived the significance of this chance finding: the sulfhydryl groups of the denatured enzyme became oxidized by air, and the enzyme spontaneously refolded into a catalytically active form. Detailed studies then showed that nearly all the original enzymatic activity was regained if the sulfhydryl groups were oxidized under suitable conditions.

All the measured physical and chemical properties of the refolded enzyme were virtually identical with those of the native enzyme. These experiments showed that *the information needed to specify the catalytically active structure of ribonuclease is contained in its amino acid sequence.* Subsequent studies have established the generality of this central principle of biochemistry: *sequence specifies conformation.* The dependence of conformation on sequence is especially significant because of the intimate connection between conformation and function.

A quite different result was obtained when reduced ribonuclease was reoxidized while it was still in 8 M urea and the preparation was then dialyzed to remove the urea. Ribonuclease reoxidized in this way had only 1per cent of the enzymatic activity of the native protein. Why were the outcomes so different when reduced ribonuclease was reoxidized in the presence and absence of urea? The reason is that the wrong disulfides formed pairs in urea. There are 105 different ways of pairing eight cysteine molecules to form four disulfides; only one of these combinations is enzymatically active. The 104 wrong pairings have been picturesquely termed "scrambled" ribonuclease. Anfinsen found that scrambled ribonuclease spontaneously converted into fully active, native ribonuclease when trace amounts of β-mercaptoethanol were added to an aqueous solution of the protein. The added β-mercaptoethanol catalyzed the rearrangement of disulfide pairings until the native structure was regained in about 10 hours. *This process was driven by the decrease in free energy as the scrambled conformations were converted into the stable, native conformation of the enzyme.* The native disulfide pairings of ribonuclease thus contribute to the stabilization of the thermodynamically preferred structure.

Similar refolding experiments have been performed on many other proteins. In many cases, the native structure can be generated under suitable conditions. For other proteins, however, refolding does not proceed efficiently. In these cases, the unfolding protein molecules usually become tangled up with one another to form aggregates. Inside cells, proteins called *chaperones* block such illicit interactions.

AMINO ACIDS HAVE DIFFERENT PROPENSITIES FOR FORMING ALPHA HELICES, BETA SHEETS, AND BETA TURNS

How does the amino acid sequence of a protein specify its three-dimensional structure? How does an unfolded polypeptide chain acquire the form of the native protein? These fundamental questions in biochemistry can be approached by first asking a simpler one: What determines whether a particular sequence in a protein forms an α helix, a β strand, or a turn? Examining the frequency of occurrence of particular amino acid residues in these secondary structures can be a source of insight into this determination. Residues such as alanine, glutamate, and leucine tend to be present in α helices, whereas valine and isoleucine tend to be present in β strands. Glycine, asparagine, and proline have a propensity for being in turns.

The results of studies of proteins and synthetic peptides have revealed some reasons for these preferences. The α helix can be regarded as the default conformation. Branching at the β-carbon atom, as in valine, threonine, and isoleucine, tends to destabilize α helices because of steric clashes. These residues are readily accommodated in β strands, in which their side chains project out of the plane containing the main chain. Serine, aspartate, and asparagine tend to disrupt α helices because their side chains contain hydrogen-bond donors or acceptors in close proximity to the main chain, where they compete for main-chain NH and CO groups. Proline tends to disrupt both α helices and β strands because it lacks an NH group and because its ring structure restricts its φ value to near -60 degrees. Glycine readily fits into all structures and for that reason does not favour helix formation in particular.

Can one predict the secondary structure of proteins by using this knowledge of the conformational preferences of amino acid residues? Predictions of secondary structure adopted by a stretch of six or fewer residues have proved to be about 60 to 70per cent accurate. What stands in the way of more accurate prediction? Note that the conformational preferences of amino acid residues are not tipped all the way to one structure. For example, glutamate, one of the strongest helix formers, prefers α helix to β strand by only a factor of two. The preference ratios of most other residues are smaller. Indeed, some penta- and hexapeptide sequences have been found to adopt one structure in one protein and an entirely different structure in another. Hence, some amino acid sequences do not uniquely determine secondary structure. Tertiary interactions—interactions between residues that are far apart in the sequence—may be decisive in specifying the secondary structure of some segments. *The context is often crucial in determining the conformational outcome.* The conformation of a protein evolved to work in a particular environment or context.

Pathological conditions can result if a protein assumes an inappropriate conformation for the context. Striking examples are *prion diseases,* such as Creutzfeldt-Jacob disease, kuru, and mad cow disease. These conditions result when a brain protein called a prion converts from its normal conformation (designated PrP^{C}) to an altered one (PrP^{Sc}). This conversion is self-propagating, leading to large aggregates of PrP^{Sc}. The role of these aggregates in the generation of the pathological conditions is not yet understood.

PROTEIN FOLDING IS A HIGHLY COOPERATIVE PROCESS

As stated earlier, proteins can be denatured by heat or by chemical denaturants such as urea or guanidium chloride. For many proteins, a comparison of the degree of unfolding as the concentration of denaturant increases has revealed a relatively sharp transition from the folded, or native, form to the unfolded, or denatured, form, suggesting that only these two conformational states are present to any significant extent. A similar sharp transition is observed if one starts with unfolded proteins and removes the denaturants, allowing the proteins to fold.

Protein folding and unfolding is thus largely an *"all or none" process* that results from a *cooperative transition.* For example, suppose that a protein is placed in conditions under which some part of the protein structure is thermodynamically unstable. As this part of the folded structure is disrupted, the interactions between it and the remainder of the protein will be lost. The loss of these interactions, in turn, will destabilize the remainder of the structure. Thus, conditions that lead to the disruption of any part of a protein structure are likely to unravel the protein completely. The structural properties of proteins provide a clear rationale for the cooperative transition.

The consequences of cooperative folding can be illustrated by considering the contents of a protein solution under conditions corresponding to the middle of the transition between the folded and unfolded forms. Under these conditions, the protein is "half folded." Yet the solution will contain no half-folded molecules but, instead, will be a 50/50 mixture of fully folded and fully unfolded molecules. Structures that are partly intact and partly disrupted are not thermodynamically stable and exist only transiently. Cooperative folding ensures that partly folded structures that might interfere with processes within cells do not accumulate.

PROTEINS FOLD BY PROGRESSIVE STABILIZATION OF INTERMEDIATES RATHER THAN BY RANDOM SEARCH

The cooperative folding of proteins is a thermodynamic property; its occurrence reveals nothing about the kinetics and mechanism of protein folding. How does a protein make the transition from a diverse ensemble of unfolded structures into a unique conformation in the native form? One possibility a priori would be that all possible conformations are tried out to find the energetically most favourable one. How long would such a random search take? Consider a small protein with 100 residues. Cyrus Levinthal calculated that, if each residue can assume three different conformations, the total number of structures would be 3^{100}, which is equal to 5×10^{47}. If it takes 10^{-13} s to convert one structure into another, the total search time would be $5 \times 10^{47} \times 10^{-13}$ s, which is equal to 5×10^{34} s, or 1.6×10^{27} years. Clearly, it would take much too long for even a small protein to fold properly by randomly trying out all possible conformations. The enormous difference between calculated and actual folding times is called *Levinthal's paradox*.

The way out of this dilemma is to recognize the power of *cumulative selection*. Richard Dawkins, in *The Blind Watchmaker*, asked how long it would take a monkey poking randomly at a typewriter to reproduce Hamlet's remark to Polonius, "Methinks it is like a weasel". An astronomically large number of keystrokes, of the order of 10^{40}, would be required. However, suppose that we preserved each correct character and allowed the monkey to retype only the wrong ones. In this case, only a few thousand keystrokes, on average, would be needed. The crucial difference between these cases is that the first employs a completely random search, whereas, in the second, *partly correct intermediates are retained*.

The essence of protein folding is the retention of partly correct intermediates. However, the protein-folding problem is much more difficult than the one presented to our simian Shakespeare. First, the criterion of correctness is not a residue-by-residue scrutiny of conformation by an omniscient observer but rather the total free energy of the transient species. Second, proteins are only marginally stable. The free-energy difference between the folded and the unfolded states of a typical 100-residue protein is 10 kcal mol^{-1} (42 kJ mol^{-1}), and thus each residue contributes on average only 0.1 kcal mol^{-1} (0.42 kJ mol^{-1}) of energy to maintain the folded state. This amount is less than that of thermal energy, which is 0.6 kcal mol^{-1} (2.5 kJ mol^{-1}) at room temperature. This meager stabilization energy means that correct intermediates, especially those formed early in folding, can be lost. The analogy is that the monkey would be somewhat free to undo its correct keystrokes. Nonetheless, the interactions that lead to cooperative folding can stabilize intermediates as structure builds up. Thus, local regions, which have significant structural preference, though not necessarily stable on their own, will tend to adopt their favoured structures and, as they form, can interact with one other, leading to increasing stabilization.

PREDICTION OF THREE-DIMENSIONAL STRUCTURE FROM SEQUENCE REMAINS A GREAT CHALLENGE

The amino acid sequence completely determines the three-dimensional structure of a protein. However, the prediction of three-dimensional structure from sequence has proved to be extremely difficult. As we have seen, the local sequence appears to determine only between 60per cent and 70per cent of the secondary structure; long-range interactions are required to fix the full secondary structure and the tertiary structure.

Investigators are exploring two fundamentally different approaches to predicting three-dimensional structure from amino acid sequence. The first is *ab initio prediction,* which attempts to predict the folding of an amino acid sequence without any direct reference to other known protein structures. Computer-based calculations are employed that attempt to minimize the free energy of a structure with a given amino acid sequence or to simulate the folding process. The utility of these methods is limited by the vast number of possible conformations, the marginal stability of proteins, and the subtle energetics of weak interactions in aqueous solution. The second approach takes advantage of our growing knowledge of the three-dimensional structures of many proteins. In these *knowledge-based methods,* an amino acid sequence of unknown structure is examined for compatibility with any known protein structures. If a significant match is detected, the known structure can be used as an initial model. Knowledge-based methods have been a source of many insights into the three-dimensional conformation of proteins of known sequence but unknown structure.

PROTEIN MODIFICATION AND CLEAVAGE CONFER NEW CAPABILITIES

Proteins are able to perform numerous functions relying solely on the versatility of their 20 amino acids. However, many proteins are covalently modifed, through the attachment of groups other than amino acids, to augment their functions. For example, *acetyl groups* are attached to the amino termini of many proteins, a modification that makes these proteins more resistant to degradation. The addition of *hy-droxyl groups* to many proline residues stabilizes fibers of newly synthesized collagen, a fibrous protein found in connective tissue and bone. The biological significance of this modification is evident in the disease scurvy: a deficiency of vitamin C results in insufficient hydroxylation of collagen and the abnormal collagen fibers that result are unable to maintain normal tissue strength. Another specialized amino acid produced by a finishing touch is *γ-carboxyglutamate.* In vitamin K deficiency, insufficient carboxylation of glutamate in prothrombin, a clotting protein, can lead to hemorrhage. Many proteins, especially those that are present on the surfaces of cells or are secreted, acquire *carbohydrate units* on specific asparagine residues. The addition of sugars makes the proteins more hydrophilic and able to participate in interactions with other proteins. Conversely, the addition of a *fatty acid* to an α-amino group or a cysteine sulfhydryl group produces a more hydrophobic protein.

Many hormones, such as epinephrine (adrenaline), alter the activities of enzymes by stimulating the phosphorylation of the hydroxyl amino acids serine and threonine; *phosphoserine* and *phosphothreonine* are the most ubiquitous modified amino acids in proteins. Growth factors such as insulin act by triggering the phosphorylation of the hydroxyl group of tyrosine residues to form *phosphotyrosine.* The phosphoryl groups on these three modified amino acids are readily removed; thus they are able to act as reversible switches in regulating cellular processes.

The preceding modifications consist of the addition of special groups to amino acids. Other special groups are generated by chemical rearrangements of side chains and, sometimes, the peptide backbone. For example, certain jellyfish produce a fluorescent green protein. The source of the fluorescence is a group formed by the spontaneous rearrangement and oxidation of the sequence Ser-Tyr-Gly within the centre of the protein. This protein is of great utility to researchers as a marker within cells.

Finally, many proteins are cleaved and trimmed after synthesis. For example, digestive enzymes are synthesized as inactive precursors that can be stored safely in the pancreas. After release into the intestine, these precursors become activated by peptide-bond cleavage. In blood clotting, peptide-bond cleavage converts soluble fibrinogen into insoluble fibrin. A number of polypeptide hormones, such as adrenocorticotropic hormone, arise from the splitting of a single large precursor protein. Likewise, many virus proteins are produced by the cleavage of large polyprotein precursors. We shall encounter many more examples of modification and cleavage as essential features of protein formation and function. Indeed, these finishing touches account for much of the versatility, precision, and elegance of protein action and regulation.

Chapter 5

Exploring Proteins

In the preceding chapter, we saw that proteins play crucial roles in nearly all biological processes—in catalysis, signal transmission, and structural support. This remarkable range of functions arises from the existence of thousands of proteins, each folded into a distinctive three-dimensional structure that enables it to interact with one or more of a highly diverse array of molecules. A major goal of biochemistry is to determine how amino acid sequences specify the conformations of proteins. Other goals are to learn how individual proteins bind specific substrates and other molecules, mediate catalysis, and transduce energy and information.

The purification of the protein of interest is the indispensable first step in a series of studies aimed at exploring protein function. Proteins can be separated from one another on the basis of solubility, size, charge, and binding ability. When a protein has been purified, the amino acid sequence can be determined. The strategy is to divide and conquer, to obtain specific fragments that can be readily sequenced. Automated peptide sequencing and the application of recombinant DNA methods are providing a wealth of amino acid sequence data that are opening new vistas. To understand the physiological context of a protein, antibodies are choice probes for locating proteins in vivo and measuring their quantities. Monoclonal antibodies able to probe for specific proteins can be obtained in large amounts. The synthesis of peptides is possible, which makes feasible the synthesis of new drugs, functional protein fragments, and antigens for inducing the formation of specific antibodies. Nuclear magnetic resonance (NMR) spectroscopy and x-ray crystallography are the principal techniques for elucidating three-dimensional structure, the key determinant of function.

The exploration of proteins by this array of physical and chemical techniques has greatly enriched our understanding of the molecular basis of life and makes it possible to tackle some of the most challenging questions of biology in molecular terms.

THE PROTEOME IS THE FUNCTIONAL REPRESENTATION OF THE GENOME

Many organisms are yielding their DNA base sequences to gene sequencers, including several metazoans. The roundworm *Caenorhabditis elegans* has a genome of 97 million bases and about 19,000 protein-encoding genes, whereas that of the fruit fly *Drosophilia melanogaster* contains 180 million bases and about 14,000 genes. The incredible progress being made in gene sequencing has already culminated in the elucidation of the complete sequence of the human genome, all 3 billion bases with an estimated 40,000 genes. But this genomic knowledge is analogous to a list of parts for a car—it does not explain how the parts work together. A new word has been coined, the *proteome,* to signify a more complex level of information content, the level of *functional information,* which encompasses the type, functions, and interactions of proteins that yield a functional unit.

The term proteome is derived from *prote*ins expressed by the gen*ome.* Whereas the genome tells us what is possible, the proteome tells us what is functionally present—for example, which proteins interact to form a signal-transduction pathway or an ion channel in a membrane. The proteome is not a fixed characteristic of the cell. Rather, because it represents the functional expression of information, it varies with cell type, developmental stage, and environmental conditions, such as the presence of hormones. The proteome is much larger than the genome because of such factors as alternatively spliced RNA, the posttranslational modification of proteins, the temporal regulation of protein synthesis, and varying protein-protein interactions. Unlike the genome, the proteome is not static.

An understanding of the proteome is acquired by investigating, characterizing, and cataloging proteins. An investigator often begins the process by separating a particular protein from all other biomolecules in the cell.

THE PURIFICATION OF PROTEINS IS AN ESSENTIAL FIRST STEP IN UNDERSTANDING THEIR FUNCTION

An adage of biochemistry is, Never waste pure thoughts on an impure protein. Starting from pure proteins, we can determine amino acid sequences and evolutionary relationships between proteins in diverse organisms and we can investigate a protein's biochemical function. Moreover, crystals of the protein may be grown from pure protein, and from such crystals we can obtain x-ray data that will provide us with a picture of the protein's tertiary structure—the actual *functional* unit.

THE ASSAY: HOW DO WE RECOGNIZE THE PROTEIN THAT WE ARE LOOKING FOR?

Purification should yield a sample of protein containing only one type of molecule, the protein in which the biochemist is interested. This protein sample may be only a fraction of 1per cent of the starting material, whether that starting material consists of cells in culture or a particular organ from a plant or animal. How is the biochemist able to isolate a particular protein from a complex mixture of proteins?

The biochemist needs a test, called an *assay*, for some unique identifying property of the protein so that he or she can tell when the protein is present. Determining an effective assay is often difficult; but the more specific the assay, the more effective the purification. For enzymes, which are protein catalysts the assay is usually based on the reaction that the enzyme catalyzes in the cell. Consider the enzyme lactate dehydrogenase, an important player in the anaerobic generation of energy from glucose as well as in the synthesis of glucose from lactate. Lactate dehydrogenase carries out the following reaction:

$$\text{Lactate} + NAD^{+} \xrightleftharpoons{\text{Lactate dehydrogenase}} \text{Pyruvate} + NADH + H^{+}$$

Nicotinamide adenine dinucleotide is distinguishable from the other components of the reaction by its ability to absorb light at 340 nm. Consequently, we can follow the progress of the reaction by examining how much light the reaction mixture absorbs at 340 nm in unit time—for instance, within 1 minute after the addition of the enzyme. Our assay for enzyme activity during the purification of lactate dehydrogenase is thus the increase in absorbance of light at 340 nm observed in 1 minute.

To be certain that our purification scheme is working, we need one additional piece of information—the amount of protein present in the mixture being assayed. There are various rapid and accurate means of determining protein concentration. With these two experimentally determined numbers—enzyme activity and protein concentration—we then calculate the *specific activity,* the ratio of enzyme activity to the amount of protein in the enzyme assay. The specific activity will rise as the purification proceeds and the protein mixture being assayed consists to a greater and greater extent of lactate dehydrogenase. In essence, the point of the purification is to maximize the specific activity.

PROTEINS MUST BE RELEASED FROM THE CELL TO BE PURIFIED

Having found an assay and chosen a source of protein, we must now fractionate the cell into components and determine which component is enriched in the protein of interest. Such fractionation schemes are developed by trial and error, on the basis of previous experience. In the first step, a *homogenate* is formed by disrupting the cell membrane, and the mixture is fractionated by centrifugation, yielding a dense pellet of heavy material at the bottom of the centrifuge tube and a lighter supernatant above. The supernatant is again centrifuged at a greater force to yield yet another pellet and supernatant. The procedure, called *differential centrifugation*, yields several fractions of decreasing density, each still containing hundreds of different proteins, which are subsequently assayed for the activity being purified. Usually, one fraction will be enriched for such activity, and it then serves as the source of material to which more discriminating purification techniques are applied.

PROTEINS CAN BE PURIFIED ACCORDING TO SOLUBILITY, SIZE, CHARGE, AND BINDING AFFINITY

Several thousand proteins have been purified in active form on the basis of such characteristics as *solubility, size, charge,* and *specific binding affinity*. Usually, protein mixtures are subjected to a series of separations, each based on a different property to yield a pure protein. At each step in the purification, the preparation is assayed and the protein concentration is determined. Substantial quantities of purified proteins, of the order of many milligrams, are needed to fully elucidate their three-dimensional structures and their mechanisms of action. Thus, the overall yield is an important feature of a purification scheme. A variety of purification techniques are available.

SALTING OUT

Most proteins are less soluble at high salt concentrations, an effect called *salting out*. The salt concentration at which a protein precipitates differs from one protein to another. Hence, salting out can be used to fractionate proteins. For example, 0.8 M ammonium sulfate precipitates fibrinogen, a blood-clotting protein, whereas a concentration of 2.4 M is needed to precipitate serum albumin. Salting out is also useful for concentrating dilute solutions of proteins, including active fractions obtained from other purification steps. Dialysis can be used to remove the salt if necessary.

DIALYSIS

Proteins can be separated from small molecules by *dialysis* through a semipermeable membrane, such as a cellulose membrane with pores. Molecules having dimensions significantly greater than the pore diameter are retained inside the dialysis bag, whereas smaller molecules and ions traverse the pores of such a membrane and emerge in the dialysate outside the bag. This technique is useful for removing a salt or other small molecule, but it will not distinguish between proteins effectively.

GEL-FILTRATION CHROMATOGRAPHY.

More discriminating separations on the basis of size can be achieved by the technique of *gel-filtration chromatography*. The sample is applied to the top of a column consisting of porous beads made of an insoluble but highly hydrated polymer such as dextran or agarose (which are carbohydrates) or polyacrylamide. Sephadex, Sepharose, and Bio-gel are commonly used commercial preparations of these beads, which are typically 100 μm (0.1 mm) in diameter. Small molecules can enter these beads, but large ones cannot. The result is that small molecules are distributed in the aqueous solution both inside the beads and between them, whereas large molecules are located only in the solution between the beads. *Large molecules flow more rapidly through this column and emerge first because a smaller volume is accessible to them.* Molecules that are of a size to occasionally enter a bead will flow from the column at an intermediate position, and small molecules, which take a longer, tortuous path, will exit last.

ION-EXCHANGE CHROMATOGRAPHY

Proteins can be separated on the basis of their net charge by *ion-exchange chromatography*. If a protein has a net positive charge at pH 7, it will usually bind to a column of beads containing carboxylate groups, whereas a negatively charged protein will not. A positively charged protein bound to such a column can then be eluted (released) by increasing the concentration of sodium chloride or another salt in the eluting buffer because sodium ions compete with positively charged groups on the protein for binding to the column. Proteins that have a low density of net positive charge will tend to emerge first, followed by those having a higher charge density. Positively charged proteins (cationic proteins) can be separated on negatively charged carboxymethyl-cellulose (CM-cellulose) columns. Conversely, negatively charged proteins (anionic proteins) can be separated by chromatography on positively charged diethylaminoethyl-cellulose (DEAE-cellulose) columns.

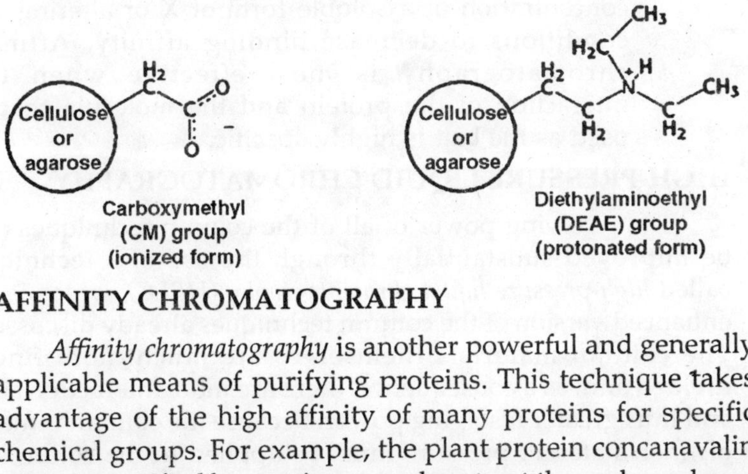

Cellulose or agarose

Carboxymethyl (CM) group (ionized form)

Cellulose or agarose

Diethylaminoethyl (DEAE) group (protonated form)

AFFINITY CHROMATOGRAPHY

Affinity chromatography is another powerful and generally applicable means of purifying proteins. This technique takes advantage of the high affinity of many proteins for specific chemical groups. For example, the plant protein concanavalin A can be purified by passing a crude extract through a column of beads containing covalently attached glucose residues. Concanavalin A binds to such a column because it has affinity for glucose, whereas most other proteins do not. The bound concanavalin A can then be released from the column by adding a concentrated solution of glucose. The glucose in solution displaces the column-attached glucose residues from binding sites on concanavalin A. Affinity chromatography is a powerful means of isolating transcription factors, proteins that regulate gene expression by binding to specific DNA sequences. A protein mixture is percolated through a column containing specific DNA sequences attached to a matrix; proteins with a high affinity for the sequence will bind and be retained. In this instance, the transcription factor is released by washing with a solution containing a high concentration of salt. In general, affinity chromatography can be effectively used to isolate a protein that recognizes group X by

1. Covalently attaching X or a derivative of it to a column,
2. Adding a mixture of proteins to this column, which is then washed with buffer to remove unbound proteins, and
3. Eluting the desired protein by adding a high concentration of a soluble form of X or altering the conditions to decrease binding affinity. Affinity chromatography is most effective when the interaction of the protein and the molecule that is used as the bait is highly specific.

HIGH-PRESSURE LIQUID CHROMATOGRAPHY

The resolving power of all of the column techniques can be improved substantially through the use of a technique called *high-pressure liquid chromatography (HPLC)*, which is an enhanced version of the column techniques already discussed. The column materials themselves are much more finely divided and, as a consequence, there are more interaction sites and thus greater resolving power. Because the column is made of finer material, pressure must be applied to the column to obtain adequate flow rates. The net result is high resolution as well as rapid separation.

PROTEINS CAN BE SEPARATED BY GEL ELECTROPHORESIS AND DISPLAYED

How can we tell whether a purification scheme is effective? One way is to ascertain that the specific activity rises with each purification step. Another is to visualize the effectiveness by displaying the proteins present at each step. The technique of electrophoresis makes the latter method possible.

GEL ELECTROPHORESIS

A molecule with a net charge will move in an electric field. This phenomenon, termed *electrophoresis*, offers a powerful means of separating proteins and other macromolecules, such as DNA and RNA. The velocity of migration *(v)* of a protein (or any molecule) in an electric field depends on the electric field strength *(E)*, the net charge on the protein *(z)*, and the frictional coefficient *(f)*.

$$V = \frac{Ea}{f} \tag{1}$$

The electric force *Ez* driving the charged molecule toward the oppositely charged electrode is opposed by the viscous drag *fv* arising from friction between the moving molecule and the medium. The frictional coefficient *f* depends on both the mass and shape of the migrating molecule and the viscosity (η) of the medium. For a sphere of radius *r*,

$$f = 6\pi\eta\rho \tag{2}$$

Electrophoretic separations are nearly always carried out in gels (or on solid supports such as paper) because the gel serves as a molecular sieve that enhances separation. Molecules that are small compared with the pores in the gel readily move through the gel, whereas molecules much larger than the pores are almost immobile. Intermediate-size molecules move through the gel with various degrees of facility. Electrophoresis is performed in a thin, vertical slab of polyacrylamide. The direction of flow is from top to bottom. Polyacrylamide gels, formed by the polymerization of acrylamide and cross-linked by methylenebisacrylamide, are choice supporting media for electrophoresis because they are chemically inert and are readily formed. Electrophoresis is the opposite of gel filtration in that all of the molecules, regardless of size, are forced to move through the same matrix. The gel behaves as one bead of a gel-filtration column.

Proteins can be separated largely on the basis of mass by electrophoresis in a polyacrylamide gel under denaturing conditions. The mixture of proteins is first dissolved in a solution of sodium dodecyl sulfate (SDS), an anionic detergent that disrupts nearly all noncovalent interactions in native proteins. Mercaptoethanol (2-thioethanol) or dithiothreitol also is added to reduce disulfide bonds. Anions of SDS bind to main chains at a ratio of about one SDS anion for every two amino acid residues. This complex of SDS with a denatured protein has a large net negative charge that is roughly proportional to the mass of the protein. The negative charge acquired on binding SDS is usually much greater than the charge on the native protein; this native charge is thus rendered insignificant. The SDS-protein complexes are then subjected to electrophoresis. When the electrophoresis is complete, the proteins in the gel can be visualized by staining them with silver or a dye such as Coomassie blue, which reveals a series of bands. Radioactive labels can be detected by placing a sheet of x-ray film over the gel, a procedure called *autoradiography*.

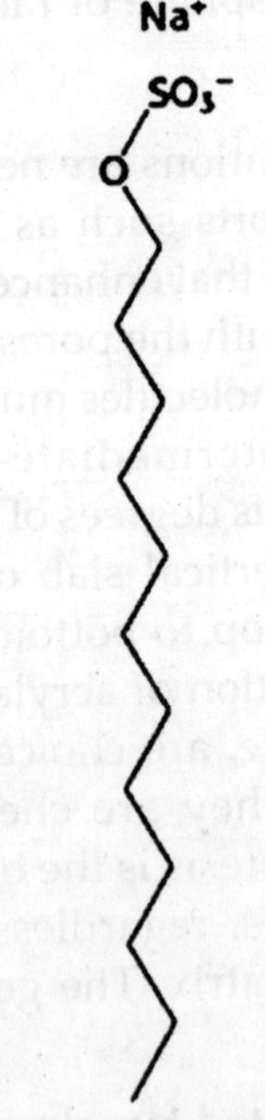

Sodium dodecyl sulfate (SDS)

Small proteins move rapidly through the gel, whereas large proteins stay at the top, near the point of application of the mixture.

The mobility of most polypeptide chains under these conditions is linearly proportional to the logarithm of their mass. Some carbohydrate-rich proteins and membrane proteins do not obey this empirical relation, however. SDS-polyacrylamide gel electrophoresis (SDS-PAGE) is rapid, sensitive, and capable of a high degree of resolution. As little as 0.1 μg (~2 pmol) of a protein gives a distinct band when stained with Coomassie blue, and even less (~0.02 μg) can be detected with a silver stain. Proteins that differ in mass by about 2per cent (e.g., 40 and 41 kd, arising from a difference of about 10 residues) can usually be distinguished.

We can examine the efficacy of our purification scheme by analyzing a part of each fraction by SDS-PAGE. The initial fractions will display dozens to hundreds of proteins. As the purification progresses, the number of bands will diminish, and the prominence of one of the bands should increase. This band will correspond to the protein of interest.

ISOELECTRIC FOCUSING

Proteins can also be separated electrophoretically on the basis of their relative contents of acidic and basic residues. The *isoelectric point* (pl) of a protein is the pH at which its net charge is zero. At this pH, its electrophoretic mobility is zero because z in equation 1 is equal to zero. For example, the pI of cytochrome *c*, a highly basic electron-transport protein, is 10.6, whereas that of serum albumin, an acidic protein in blood, is 4.8. Suppose that a mixture of proteins undergoes electrophoresis in a pH gradient in a gel in the absence of SDS. Each protein will move until it reaches a position in the gel at which the pH is equal to the pI of the protein. This method of separating proteins according to their isoelectric point is called *isoelectric focusing*. The pH gradient in the gel is formed first by subjecting a mixture of *polyampholytes* (small multicharged polymers) having many pI values to electrophoresis. Isoelectric focusing can readily resolve proteins that differ in pI by as little as 0.01, which means that proteins differing by one net charge can be separated.

TWO-DIMENSIONAL ELECTROPHORESIS

Isoelectric focusing can be combined with SDS-PAGE to obtain very high resolution separations. A single sample is first subjected to isoelectric focusing. This single-lane gel is then placed horizontally on top of an SDS-polyacrylamide slab. The proteins are thus spread across the top of the polyacrylamide gel according to how far they migrated during isoelectric focusing. They then undergo electrophoresis again in a perpendicular direction (vertically) to yield a twodimensional pattern of spots. In such a gel, proteins have been separated in the horizontal direction on the basis of isoelectric point and in the vertical direction on the basis of mass. It is remarkable that more than a thousand different proteins in the bacterium *Escherichia coli* can be resolved in a single experiment by two-dimensional electrophoresis.

Proteins isolated from cells under different physiological conditions can be subjected to two-dimensional electrophoresis, followed by an examination of the intensity of the signals. In this way, particular proteins can be seen to increase or decrease in concentration in response to the physiological state. How can we tell what protein is being regulated? A former drawback to the power of the two-dimensional gel is that, although many proteins are displayed, they are not identified. It is now possible to identify proteins by coupling two-dimensional gel electrophoresis with mass spectrometric techniques. We will consider these techniques when we examine how the mass of a protein is determined.

A PROTEIN PURIFICATION SCHEME CAN BE QUANTITATIVELY EVALUATED

To determine the success of a protein purification scheme, we monitor the procedure at each step by determining specific activity and by performing an SDS-PAGE analysis. Consider the results for the purification of a fictitious protein. At each step, the following parameters are measured:

Total Protein

The quantity of protein present in a fraction is obtained by determining the protein concentration of a part of each fraction and multiplying by the fraction's total volume.

Total Activity

The enzyme activity for the fraction is obtained by measuring the enzyme activity in the volume of fraction used in the assay and multiplying by the fraction's total volume.

Specific Activity

This parameter is obtained by dividing total activity by total protein.

Yield. This parameter is a measure of the activity retained after each purification step as a percentage of the activity in the crude extract. The amount of activity in the initial extract is taken to be 100per cent.

Purification Level

This parameter is a measure of the increase in purity and is obtained by dividing the specific activity, calculated after each purification step, by the specific activity of the initial extract.

As we see the first purification step, salt fractionation, leads to an increase in purity of only 3-fold, but we recover nearly all the target protein in the original extract, given that the yield is 92per cent. After dialysis to lower the high concentration of salt remaining from the salt fractionation, the fraction is passed through an ion-exchange column. The purification now increases to 9-fold compared with the original extract, whereas the yield falls to 77per cent. Molecular exclusion chromatography brings the level of purification to 100-fold, but the yield is now at 50per cent. The final step is affinity chromatography with the use of a ligand specific for the target enzyme. This step, the most powerful of these purification procedures, results in a purification level of 3000-fold, while lowering the yield to 35per cent. If we load a constant amount of protein onto each lane after each step, the number of bands decreases in proportion to the level of purification, and the amount of protein of interest increases as a proportion of the total protein present.

A good purification scheme takes into account both purification levels and yield. A high degree of purification and a poor yield leave little protein with which to experiment. A high yield with low purification leaves many contaminants (proteins other than the one of interest) in the fraction and complicates the interpretation of experiments.

ULTRACENTRIFUGATION IS VALUABLE FOR SEPARATING BIOMOLECULES AND DETERMINING THEIR MASSES

We have already seen that centrifugation is a powerful and generally applicable method for separating a crude mixture of cell components, but it is also useful for separating and analyzing biomolecules themselves. With this technique, we can determine such parameters as mass and density, learn something about the shape of a molecule, and investigate the interactions between molecules. To deduce these properties from the centrifugation data, we need a mathematical description of how a particle behaves in a centrifugal force.

A particle will move through a liquid medium when subjected to a centrifugal force. A convenient means of quantifying the rate of movement is to calculate the sedimentation coefficient, *s*, of a particle by using the following equation:

$$s = m(1 - \bar{v}\rho)/f$$

where *m* is the mass of the particle, v^{-} is the partial specific volume (the reciprocal of the particle density), ρ is the density of the medium and *f* is the frictional coefficient (a measure of the shape of the particle). The (1 - ρ) term is the buoyant force exerted by liquid medium.

Sedimentation coefficients are usually expressed in *Svedberg units (S)*, equal to 10^{-13} s. The smaller the S value, the slower a molecule moves in a centrifugal field.

Several important conclusions can be drawn from the preceding equation:

1. The sedimentation velocity of a particle depends in part on its mass. A more massive particle sediments more rapidly than does a less massive particle of the same shape and density.
2. Shape, too, influences the sedimentation velocity because it affects the viscous drag. The frictional coefficient f of a compact particle is smaller than that of an extended particle of the same mass. Hence, elongated particles sediment more slowly than do spherical ones of the same mass.
3. A dense particle moves more rapidly than does a less dense one because the opposing buoyant force $(1 - \rho)$ is smaller for the denser particle.
4. The sedimentation velocity also depends on the density of the solution. (ρ). Particles sink when $\rho < 1$, float when $\rho > 1$, and do not move when $\rho = 1$.

A technique called *zonal, band,* or most commonly *gradient* centrifugation can be used to separate proteins with different sedimentation coefficients. The first step is to form a density gradient in a centrifuge tube. Differing proportions of a low-density solution (such as 5per cent sucrose) and a high-density solution (such as 20per cent sucrose) are mixed to create a linear gradient of sucrose concentration ranging from 20per cent at the bottom of the tube to 5per cent at the top. The role of the gradient is to prevent connective flow. A small volume of a solution containing the mixture of proteins to be separated is placed on top of the density gradient. When the rotor is spun, proteins move through the gradient and separate according to their sedimentation coefficients. The time and speed of the centrifugation is determined empirically. The separated bands, or zones, of protein can be harvested by making a hole in the bottom of the tube and collecting drops. The drops can be measured for protein content and catalytic activity or another functional property. This sedimentation-velocity technique readily separates proteins differing in sedimentation coefficient by a factor of two or more.

The mass of a protein can be directly determined by *sedimentation equilibrium,* in which a sample is centrifuged at relatively low speed so that sedimentation is counterbalanced by diffusion. *The sedimentation-equilibrium technique for determining mass is very accurate and can be applied under nondenaturing conditions in which the native quaternary structure of multimeric proteins is preserved.* In contrast, SDS-polyacrylamide gel electrophoresis provides an *estimate* of the mass of dissociated polypeptide chains under *denaturing* conditions. Note that, if we know the mass of the dissociated components of a multimeric protein as determined by SDS-polyacrylamide analysis and the mass of the intact multimeric protein as determined by sedimentation equilibrium analysis, we can determine how many copies of each polypeptide chain is present in the multimeric protein.

THE MASS OF A PROTEIN CAN BE PRECISELY DETERMINED BY MASS SPECTROMETRY

Mass spectrometry has been an established analytical technique in organic chemistry for many years. Until recently, however, the very low volatility of proteins made mass spectrometry useless for the investigation of these molecules. This difficulty has been circumvented by the introduction of techniques for effectively dispersing proteins and other macromolecules into the gas phase. These methods are called *matrix-assisted laser desorption-ionization (MALDI)* and *electrospray spectrometry.* We will focus on MALDI spectrometry. In this technique, protein ions are generated and then accelerated through an electrical field. They travel through the flight tube, with the smallest traveling fastest and arriving at the detector first. Thus, the *time of flight (TOF)* in the electrical field is a measure of the mass (or, more precisely, the mass/charge ratio). Tiny amounts of biomolecules, as small as a few picomoles (pmol) to femtomoles (fmol), can be analysed in this manner. A MALDI-TOF mass spectrum for a mixture of the proteins insulin and β-lactoglobulin. The masses determined by MALDI-TOF are 5733.9 and 18,364, respectively, compared with calculated values of 5733.5 and

18,388. MALDI-TOF is indeed an accurate means of determining protein mass.

Mass spectrometry has permitted the development of *peptide mass fingerprinting*. This technique for identifying peptides has greatly enhanced the utility of two-dimensional gels. The sample of interest is extracted and cleaved *specifically* by chemical or enzymatic means. The masses of the protein fragments are then determined with the use of mass spectrometry. Finally, the peptide masses, or *fingerprint*, are matched against the fingerprint found in databases of proteins that have been "electronically cleaved" by a computer simulating the same fragmentation technique used for the experimental sample. This technique has provided some outstanding results. For example, of 150 yeast proteins analysed with the use of two-dimensional gels, peptide mass fingerprinting unambiguously identified 80per cent. Mass spectrometry has provided name tags for many of the proteins in twodimensional gels.

AMINO ACID SEQUENCES CAN BE DETERMINED BY AUTOMATED EDMAN DEGRADATION

The protein of interest having been purified and its mass determined, the next analysis usually performed is to determine the protein's amino acid sequence, or primary structure. As stated previously a wealth of information about a protein's function and evolutionary history can often be obtained from the primary structure. Let us examine first how we can sequence a simple peptide, such as

Ala-Gly-Asp-Phe-Arg-Gly

The first step is to determine the *amino acid composition* of the peptide. The peptide is hydrolyzed into its constituent amino acids by heating it in 6 N HCl at 110°C for 24 hours. Amino acids in hydrolysates can be separated by ion-exchange chromatography on columns of sulfonated polystyrene. The identity of the amino acid is revealed by its elution volume, which is the volume of buffer used to remove the amino acid from the column and quantified by reaction with *ninhydrin.* Amino acids treated with ninhydrin give an intense blue colour, except for proline, which gives a yellow colour because it contains a secondary amino group. The concentration of an amino acid in a solution, after heating with ninhydrin, is proportional to the optical absorbance of the solution. This technique can detect a microgram (10 nmol) of an amino acid, which is about the amount present in a thumbprint. As little as a nanogram (10 pmol) of an amino acid can be detected by replacing ninhydrin with *fluorescamine,* which reacts with the α-amino group to form a highly fluorescent product. A comparison of the chromatographic patterns of our sample hydrolysate with that of a standard mixture of amino acids would show that the amino acid composition of the peptide is

(Ala, Arg, Asp, Gly_2, Phe)

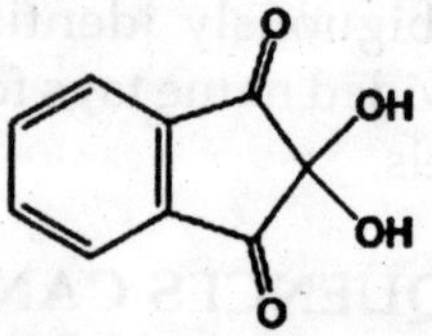

OH
OH
O
O

The parentheses denote that this is the amino acid composition of the peptide, not its sequence.

The next step is often to identify the N-terminal amino acid by labeling it with a compound that forms a stable covalent bond. *Fluorodinitrobenzene* (FDNB) was first used for this purpose by Frederick Sanger. *Dabsyl chloride* is now commonly used because it forms fluorescent derivatives that can be detected with high sensitivity. It reacts with an uncharged α-NH_2 group to form a sulfonamide derivative that is stable under conditions that hydrolyze peptide bonds. Hydrolysis of our sample dabsyl-peptide in 6 N HCl would yield a dabsyl-amino acid, which could be identified as dabsyl-alanine by its chromatographic properties. *Dansyl chloride,* too, is a valuable labeling reagent because it forms fluorescent sulfonamides.

Although the dabsyl method for determining the amino-terminal residue is sensitive and powerful, it cannot be used repeatedly on the same peptide, because the peptide is totally degraded in the acid-hydrolysis step and thus all sequence information is lost. Pehr Edman devised a method for labeling the amino-terminal residue and cleaving it from the peptide without disrupting the peptide bonds between the other amino acid residues. The *Edman degradation* sequentially removes one residue at a time from the amino end of a peptide. *Phenyl isothiocyanate* reacts with the uncharged terminal amino group of the peptide to form a phenylthiocarbamoyl derivative. Then, under mildly acidic conditions, a cyclic derivative of the terminal amino acid is liberated, which leaves an intact peptide shortened by one amino acid. The cyclic compound is a phenylthiohydantoin (PTH)-amino acid, which can be identified by chromatographic procedures. The Edman procedure can then be repeated on the shortened peptide, yielding another PTH-amino acid, which can again be identified by chromatography. Three more rounds of the Edman degradation will reveal the complete sequence of the original peptide pentapeptide.

Fluorodinitrobenzene **Dabsyl chloride**

The development of automated sequencers has markedly decreased the time required to determine protein sequences. One cycle of the Edman degradation—the cleavage of an amino acid from a peptide and its identification—is carried out in less than 1 hour. By repeated degradations, the amino acid sequence of some 50 residues in a protein can be determined. High-pressure liquid chromatography provides a sensitive means of distinguishing the various amino acids. Gas-phase sequenators can analyse picomole quantities of peptides and proteins. This high sensitivity makes it feasible to analyse the sequence of a protein sample eluted from a single band of an SDS-polyacrylamide gel.

PROTEINS CAN BE SPECIFICALLY CLEAVED INTO SMALL PEPTIDES TO FACILITATE ANALYSIS

In principle, it should be possible to sequence an entire protein by using the Edman method. In practice, the peptides cannot be much longer than about 50 residues. This is so because the reactions of the Edman method, especially the release step, are not 100per cent efficient, and so not all peptides in the reaction mixture release the amino acid derivative at each step. For instance, if the efficiency of release for each round were 98per cent, the proportion of "correct" amino acid released after 60 rounds would be (0.98^{60}), or 0.3—a hopelessly impure mix. This obstacle can be circumvented by cleaving the original protein at specific amino acids into smaller peptides that can be sequenced. In essence, the strategy is to *divide and conquer*.

Specific cleavage can be achieved by chemical or enzymatic methods. For example, *cyanogen bromide* (CNBr) splits polypeptide chains only on the carboxyl side of methionine residues.

A protein that has 10 methionine residues will usually yield 11 peptides on cleavage with CNBr. Highly specific cleavage is also obtained with *trypsin*, a proteolytic enzyme from pancreatic juice. Trypsin cleaves polypeptide chains on the carboxyl side of arginine and lysine residues protein that contains 9 lysine and 7 arginine residues will usually yield 17 peptides on digestion with trypsin. Each of these tryptic peptides, except for the carboxyl-terminal peptide of the protein, will end with either arginine or lysine.

The peptides obtained by specific chemical or enzymatic cleavage are separated by some type of chromatography. The sequence of each purified peptide is then determined by the Edman method. At this point, the amino acid sequences of segments of the protein are known, but the order of these segments is not yet defined. How can we order the peptides to obtain the primary structure of the original protein? The necessary additional information is obtained from *overlap peptides*. A second enzyme is used to split the polypeptide chain at different linkages. For example, chymotrypsin cleaves preferentially on the carboxyl side of aromatic and some other bulky nonpolar residues. Because these chymotryptic peptides overlap two or more tryptic peptides, they can be used to establish the order of the peptides. The entire amino acid sequence of the polypeptide chain is then known.

Additional steps are necessary if the initial protein sample is actually several polypeptide chains. SDS-gel electrophoresis under reducing conditions should display the number of chains. Alternatively, the number of distinct N-terminal amino acids could be determined. For a protein made up of two or more polypeptide chains held together by noncovalent bonds, denaturing agents, such as urea or guanidine hydrochloride, are used to dissociate the chains from one another. The dissociated chains must be separated from one another before sequence determination of the individual chains can begin. Polypeptide chains linked by disulfide bonds are separated by reduction with thiols such as β-mercaptoethanol or dithiothreitol. To prevent the cysteine residues from recombining, they are then alkylated with iodoacetate to form stable *S*-carboxymethyl derivatives. Sequencing can then be performed as heretofore described.

To complete our understanding of the protein's structure, we need to determine the positions of the original disulfide bonds. This information can be obtained by using a *diagonal electrophoresis* technique to isolate the peptide sequences containing such bonds. First, the protein is specifically cleaved into peptides under conditions in which the disulfides remain intact. The mixture of peptides is applied to a corner of a sheet of paper and subjected to electrophoresis in a single lane along one side. The resulting sheet is exposed to vapours of performic acid, which cleaves disulfides and converts them into cysteic acid residues. Peptides originally linked by disulfides are now independent and more acidic because of the formation of an SO_3^- group.

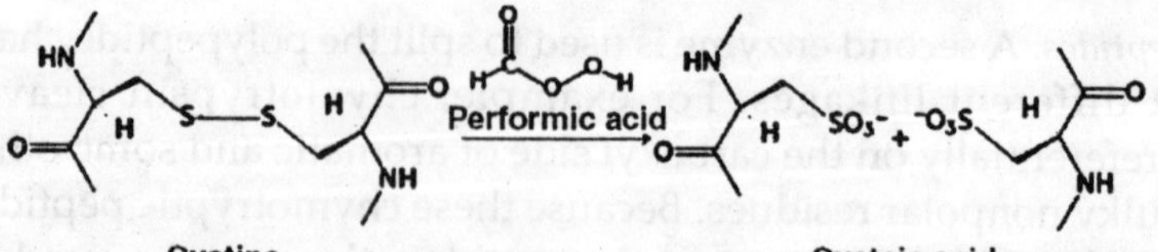

Performic acid

Cystine Cysteic acid

This mixture is subjected to electrophoresis in the perpendicular direction under the same conditions as those of the first electrophoresis. Peptides that were devoid of disulfides will have the same mobility as before, and consequently all will be located on a single diagonal line. In contrast, the newly formed peptides containing cysteic acid will usually migrate differently from their parent disulfide-linked peptides and hence will lie off the diagonal. These peptides can then be isolated and sequenced, and the location of the disulfide bond can be established.

AMINO ACID SEQUENCES ARE SOURCES OF MANY KINDS OF INSIGHT

A protein's amino acid sequence, once determined, is a valuable source of insight into the protein's function, structure, and history.

1. The sequence of a protein of interest can be compared with all other known sequences to ascertain whether significant similarities exist. Does this protein belong to one of the established families? A search for kinship between a newly sequenced protein and the thousands of previously sequenced ones takes only a few seconds on a personal computer. If the newly isolated protein is a member of one of the established classes of protein, we can begin to infer information about the protein's function. For instance, chymotrypsin and trypsin are members of the serine protease family, a clan of proteolytic enzymes that have a common catalytic mechanism based on a reactive serine residue. If the sequence of the newly isolated protein shows sequence similarity with trypsin or chymotrypsin, the result suggests that it may be a serine protease.

2. *Comparison of sequences of the same protein in different species yields a wealth of information about evolutionary pathways.* Genealogical relations between species can be inferred from sequence differences between their proteins. We can even estimate the time at which two evolutionary lines diverged, thanks to the clocklike nature of random mutations. For example, a comparison of serum albumins found in primates indicates that human beings and African apes diverged 5 million years ago, not 30 million years ago as was once thought. Sequence analyses have opened a new perspective on the fossil record and the pathway of human evolution.
3. *Amino acid sequences can be searched for the presence of internal repeats.* Such internal repeats can reveal information about the history of an individual protein itself. Many proteins apparently have arisen by duplication of a primordial gene followed by its diversification. For example, calmodulin, a ubiquitous calcium sensor in eukaryotes, contains four similar calcium-binding modules that arose by gene duplication.
4. *Many proteins contain amino acid sequences that serve as signals designating their destinations or controlling their processing.* A protein destined for export from a cell or for location in a membrane, for example, contains a signal sequence, a stretch of about 20 hydrophobic residues near the amino terminus that directs the protein to the appropriate membrane. Another protein may contain a stretch of amino acids that functions as a nuclear localization signal, directing the protein to the nucleus.
5. *Sequence data provide a basis for preparing antibodies specific for a protein of interest.* Careful examination of the amino acid sequence of a protein can reveal which sequences will be most likely to elicit an antibody when injected into a mouse or rabbit.

Peptides with these sequences can be synthesized and used to generate antibodies to the protein. These specific antibodies can be very useful in determining the amount of a protein present in solution or in the blood, ascertaining its distribution within a cell, or cloning its gene.

6. *Amino acid sequences are valuable for making DNA probes that are specific for the genes encoding the corresponding proteins.* Knowledge of a protein's primary structure permits the use of reverse genetics. DNA probes that correspond to a part of the amino acid sequence can be constructed on the basis of the genetic code. These probes can be used to isolate the gene of the protein so that the entire sequence of the protein can be determined. The gene in turn can provide valuable information about the physiological regulation of the protein. Protein sequencing is an integral part of molecular genetics, just as DNA cloning is central to the analysis of protein structure and function.

RECOMBINANT DNA TECHNOLOGY HAS REVOLUTIONIZED PROTEIN SEQUENCING

Hundreds of proteins have been sequenced by Edman degradation of peptides derived from specific cleavages. Nevertheless, heroic effort is required to elucidate the sequence of large proteins, those with more than 1000 residues. For sequencing such proteins, a complementary experimental approach based on recombinant DNA technology is often more efficient. As will be discussed in Chapter 6, long stretches of DNA can be cloned and sequenced, and the nucleotide sequence directly reveals the amino acid sequence of the protein encoded by the gene. Recombinant DNA technology is producing a wealth of amino acid sequence information at a remarkable rate.

Even with the use of the DNA base sequence to determine primary structure, there is still a need to work with isolated proteins. The amino acid sequence deduced by reading the DNA sequence is that of the *nascent* protein, the direct product of the translational machinery. Many proteins are modified after synthesis. Some have their ends trimmed, and others arise by cleavage of a larger initial polypeptide chain. Cysteine residues in some proteins are oxidized to form disulfide links, connecting either parts within a chain or separate polypeptide chains. Specific side chains of some proteins are altered. Amino acid sequences derived from DNA sequences are rich in information, but they do not disclose such posttranslational modifications. Chemical analyses of proteins in their final form are needed to delineate the nature of these changes, which are critical for the biological activities of most proteins. Thus, genomic and proteomic analyses are complementary approaches to elucidating the structural basis of protein function.

IMMUNOLOGY PROVIDES IMPORTANT TECHNIQUES WITH WHICH TO INVESTIGATE PROTEINS

Immunological methods have become important tools used to purify a protein, locate it in the cell, or quantify how much of the protein is present. These methods are predicated on the exquisite specificity of antibodies for their target proteins. Labeled antibodies provide a means to tag a specific protein so that it can be isolated, quantified, or visualized.

ANTIBODIES TO SPECIFIC PROTEINS CAN BE GENERATED

Immunological techniques begin with the generation of antibodies to a particular protein. An *antibody* (also called an *immunoglobulin,* Ig) is a protein synthesized by an animal in response to the presence of a foreign substance, called an *antigen,* and normally functions to protect the animal from infection. Antibodies have specific and high affinity for the antigens that elicited their synthesis. Proteins, polysaccharides, and nucleic acids can be effective antigens. An antibody recognizes a specific group or cluster of amino acids on a large molecule called an *antigenic determinant,* or *epitope.* Small foreign molecules, such as synthetic peptides, also can elicit antibodies, provided that the small molecule contains a recognized epitope and is attached to a macromolecular carrier. The small foreign molecule itself is called a *hapten*. Animals have a very large repertoire of antibody-producing cells, each producing an antibody of a single specificity. An antigen acts by stimulating the proliferation of the small number of cells that were already forming an antibody capable of recognizing the antigen.

Immunological techniques depend on our being able to generate antibodies to a specific antigen. To obtain antibodies that recognize a particular protein, a biochemist injects the protein into a rabbit twice, 3 weeks apart. The injected protein stimulates the reproduction of cells producing antibodies that recognize the foreign substance. Blood is drawn from the immunized rabbit several weeks later and centrifuged to separate blood cells from the supernatant, or serum. The serum, called an *antiserum,* contains antibodies to all antigens to which the rabbit has been exposed. Only some of them will be antibodies to the injected protein. Moreover, antibodies of a given specificity are not a single molecular species. For instance, 2,4- dinitrophenol (DNP) has been used as a hapten to generate antibodies to DNP. Analyses of anti-DNP antibodies revealed a wide range of binding affinities—the dissociation constants ranged from about 0.1 nM to 1 μM. Correspondingly, a large number of bands were evident when anti-DNP antibody was subjected to isoelectric focusing. These results indicate that cells are producing many different antibodies, each recognizing a different surface feature of the same antigen. The antibodies are heterogeneous, or *polyclonal*. This heterogeneity is a barrier, which can complicate the use of these antibodies.

MONOCLONAL ANTIBODIES WITH VIRTUALLY ANY DESIRED SPECIFICITY CAN BE READILY PREPARED

The discovery of a means of producing *monoclonal antibodies* of virtually any desired specificity was a major breakthrough that intensified the power of immunological approaches. Just as working with impure proteins makes it difficult to interpret data and understand function, so too does working with an impure mixture of antibodies. The ideal would be to isolate a clone of cells that produce only a single antibody. The problem is that antibody-producing cells isolated from an organism die in a short time.

Immortal cell lines that produce monoclonal antibodies do exist. These cell lines are derived from a type of cancer, *multiple myeloma,* a malignant disorder of antibody-producing cells. In this cancer, a single transformed plasma cell divides uncontrollably, generating a very large number of *cells of a single kind*. They are a *clone* because they are descended from the same cell and have identical properties. The identical cells of the myeloma secrete large amounts of normal *immunoglobulin of a single kind* generation after generation. A myeloma can be transplanted from one mouse to another, where it continues to proliferate.

Cesar Milstein and Georges Köhler discovered that *large amounts of homogeneous antibody of nearly any desired specificity could be obtained by fusing a short-lived antibody-producing cell with an immortal myeloma cell*. An antigen is injected into a mouse, and its spleen is removed several weeks later. A mixture of plasma cells from this spleen is fused in vitro with myeloma cells. Each of the resulting hybrid cells, called *hybridoma cells,* indefinitely produces homogeneous antibody specified by the parent cell from the spleen. Hybridoma cells can then be screened, by using some sort of assay for the antigen-antibody interaction, to determine which ones produce antibody having the desired specificity. Collections of cells shown to produce the desired antibody are subdivided and reassayed. This process is repeated until a pure cell line, a clone producing a single antibody, is isolated. These positive cells can be grown in culture medium or injected into mice to induce myelomas. Alternatively, the cells can be frozen and stored for long periods.

The hybridoma method of producing monoclonal antibodies has opened new vistas in biology and medicine.

Large amounts of homogeneous antibodies with tailor-made specificities can be readily prepared. They are sources of insight into relations between antibody structure and specificity. Moreover, monoclonal antibodies can serve as precise analytical and preparative reagents. For example, a pure antibody can be obtained against an antigen that has not yet been isolated. Proteins that guide development have been identified with the use of monoclonal antibodies as tags. Monoclonal antibodies attached to solid supports can be used as affinity columns to purify scarce proteins. This method has been used to purify interferon (an antiviral protein) 5000-fold from a crude mixture. *Clinical laboratories are using monoclonal antibodies in many assays.* For example, the detection in blood of isozymes that are normally localized in the heart points to a myocardial infarction (heart attack). Blood transfusions have been made safer by antibody screening of donor blood for viruses that cause AIDS (acquired immune deficiency syndrome), hepatitis, and other infectious diseases. Monoclonal antibodies are also being evaluated for use as therapeutic agents, as in the treatment of cancer. Furthermore, the vast repertoire of antibody specificity can be tapped to generate catalytic antibodies having novel features not found in naturally occurring enzymes.

PROTEINS CAN BE DETECTED AND QUANTITATED BY USING AN ENZYME-LINKED IMMUNOSORBENT ASSAY

Antibodies can be used as exquisitely specific analytic reagents to quantify the amount of a protein or other antigen. The technique is the *enzyme-linked immunosorbent assay (ELISA).* In this method, an enzyme, which reacts with a colourless substrate to produce a coloured product, is covalently linked to a specific antibody that recognizes a target antigen. If the antigen is present, the antibody-enzyme complex will bind to it, and the enzyme component of the antibody-enzyme complex will catalyze the reaction generating the coloured product. Thus, the presence of the coloured product indicates the presence of the antigen. Such an enzyme-linked immunosorbent assay, which is rapid and convenient, can detect less than a nanogram (10^{-9} g) of a protein. ELISA can be performed with either polyclonal or monoclonal antibodies, but the use of monoclonal antibodies yields more reliable results.

We will consider two among the several types of ELISA. *The indirect ELISA is used to detect the presence of antibody* and is the basis of the test for HIV infection. In that test, viral core proteins (the antigen) are absorbed to the bottom of a well. Antibodies from a patient are then added to the coated well and allowed to bind to the antigen. Finally, enzyme-linked antibodies to human antibodies (for instance, goat antibodies that recognize human antibodies) are allowed to react in the well and unbound antibodies are removed by washing. Substrate is then applied. An enzyme reaction suggests that the enzyme-linked antibodies were bound to human antibodies, which in turn implies that the patient had antibodies to the viral antigen.

The sandwich ELISA allows both the detection and the quantitation of antigen. Antibody to a particular antigen is first absorbed to the bottom of a well. Next, the antigen (or blood or urine containing the antigen) is added to the well and binds to the antibody. Finally, a second, different antibody to the antigen is added. This antibody is enzyme linked and is processed as described for indirect ELISA. In this case, the extent of reaction is directly proportional to the amount of antigen present. Consequently, it permits the measurement of small quantities of antigen.

WESTERN BLOTTING PERMITS THE DETECTION OF PROTEINS SEPARATED BY GEL ELECTROPHORESIS

Often it is necessary to detect small quantities of a particular protein in the presence of many other proteins, such as a viral protein in the blood. Very small quantities of a protein of interest in a cell or in body fluid can be detected by an immunoassay technique called *Western blotting*. A sample is subjected to electrophoresis on an SDS-polyacrylamide gel. Blotting (or more typically electroblotting) transfers the resolved proteins on the gel to the surface of a polymer sheet to make them more accessible for reaction. An antibody that is specific for the protein of interest is added to the sheet and reacts with the antigen. The antibody-antigen complex on the sheet then can be detected by rinsing the sheet with a second antibody specific for the first (e.g., goat antibody that recognizes mouse antibody). A radioactive label on the second antibody produces a dark band on x-ray film (an autoradiogram). Alternatively, an enzyme on the second antibody generates a coloured product, as in the ELISA method. Western blotting makes it possible to find a protein in a complex mixture, the proverbial needle in a haystack. It is the basis for the test for infection by hepatitis C, where it is used to detect a core protein of the virus. This technique is also very useful in the cloning of genes.

FLUORESCENT MARKERS MAKE POSSIBLE THE VISUALIZATION OF PROTEINS IN THE CELL

Biochemistry is often performed in test tubes or polyacrylamide gels. However, most proteins function in the context of a cell. Fluorescent markers provide a powerful means of examining proteins in their biological context. For instance, cells can be stained with fluorescence-labeled antibodies or other fluorescent proteins and examined by *fluorescence microscopy* to reveal the location of a protein of interest. Arrays of parallel bundles are evident in cells stained with antibody specific for actin, a protein that polymerizes into filaments. Actin filaments are constituents of the cytoskeleton, the internal scaffolding of cells that controls their shape and movement. By tracking protein location, fluorescent markers also provide clues to protein function. For instance, the glucocorticoid receptor protein is a transcription factor that controls gene expression in response to the steroid hormone cortisone. The receptor was linked to *green fluorescent protein (GPF),* a naturally fluorescent protein isolated from the jellyfish *Aequorea victoria*. Fluorescence microscopy revealed that, in the absence of the hormone, the receptor is located in the cytoplasm. On addition of the steroid, the receptor is translocated to the nucleus, where it binds to DNA.

The highest resolution of fluorescence microscopy is about 0.2 μm (200 nm, or 2000 Å), the wavelength of visible light. Finer spatial resolution can be achieved by electron microscopy by using antibodies tagged with electron-dense markers. For example, ferritin conjugated to an antibody can be readily visualized by electron microscopy because it contains an electron-dense core rich in iron. Clusters of gold also can be conjugated to antibodies to make them highly visible under the electron microscope. *Immunoelectron microscopy* can define the position of antigens to a resolution of 10 nm (100 Å) or finer.

PEPTIDES CAN BE SYNTHESIZED BY AUTOMATED SOLID-PHASE METHODS

The ability to synthesize peptides of defined sequence is a powerful technique for extending biochemical analysis for several reasons.

1. *Synthetic peptides can serve as antigens to stimulate the formation of specific antibodies.* For instance, as discussed earlier, it is often more efficient to obtain a protein sequence from a nucleic acid sequence than by sequencing the protein itself. Peptides can be synthesized on the basis of the nucleic acid sequence, and antibodies can be raised that target these peptides. These antibodies can then be used to isolate the intact protein from the cell.

2. *Synthetic peptides can be used to isolate receptors for many hormones and other signal molecules.* For example, white blood cells are attracted to bacteria by formylmethionyl (fMet) peptides released in the breakdown of bacterial proteins. Synthetic formylmethionyl peptides have been useful in identifying the cell-surface receptor for this class of peptide. Moreover, synthetic peptides can be attached to agarose beads to prepare affinity chromatography columns for the purification of receptor proteins that specifically recognize the peptides.

3. *Synthetic peptides can serve as drugs.* Vasopressin is a peptide hormone that stimulates the reabsorption of water in the distal tubules of the kidney, leading to the formation of more concentrated urine. Patients with diabetes insipidus are deficient in *vasopressin* (also called *antidiuretic hormone*), and so they excrete large volumes of urine (more than 5 liters per day) and are continually thirsty. This defect can be treated by administering 1-desamino-8-d-arginine vasopressin, a synthetic analog of the missing hormone. This synthetic peptide is degraded in vivo much more slowly than vasopressin and, additionally, does not increase the blood pressure.

4. Finally, *studying synthetic peptides can help define the rules governing the three-dimensional structure of proteins.* We can ask whether a particular sequence by itself folds into an *a* helix, *b* strand, or hairpin turn or behaves as a random coil.

How are these peptides constructed? The amino group of one amino acid is linked to the carboxyl group of another. However, a unique product is formed only if a single amino group and a single carboxyl group are available for reaction. Therefore, it is necessary to block some groups and to activate others to prevent unwanted reactions. The α-amino group of the first amino acid of the desired peptide is blocked with a *tert*-butyloxycarbonyl (*t*-Boc) group, yielding a *t*-Boc amino acid. The carboxyl group of this same amino acid is activated by reacting it with a reagent such as *dicyclohexylcarbodiimide* (DCC). The free amino group of the next amino acid to be linked attacks the activated carboxyl, leading to the formation of a peptide bond and the release of dicyclohexylurea. The carboxyl group of the resulting dipeptide is activated with DCC and reacted with the free amino group of the amino acid that will be the third residue in the peptide. This process is repeated until the desired peptide is synthesized. Exposing the peptide to dilute acid removes the *t*-Boc protecting group from the first amino acid while leaving peptide bonds intact.

Peptides containing more than 100 amino acids can be synthesized by sequential repetition of the preceding reactions. Linking the growing peptide chain to an insoluble matrix, such as polystyrene beads, further enhances efficiency. A major advantage of this *solid-phase method* is that the desired product at each stage is bound to beads that can be rapidly filtered and washed, and so there is no need to purify intermediates. All reactions are carried out in a single vessel, eliminating losses caused by repeated transfers of products. The carboxyl-terminal amino acid of the desired peptide sequence is first anchored to the polystyrene beads. The *t*-Boc protecting group of this amino acid is then removed. The next amino acid (in the protected *t*-Boc form) and dicyclohexylcarbodiimide, the coupling agent, are added together. After the peptide bond forms, excess reagents and dicyclohexylurea are washed away, leaving the desired dipeptide product attached to the beads. Additional amino acids are linked by the same sequence of reactions. At the end of the synthesis, the peptide is released from the beads by adding hydrofluoric acid (HF), which cleaves the carboxyl ester anchor without disrupting peptide bonds. Protecting groups on potentially reactive side chains, such as that of lysine, also are removed at this time. This cycle of reactions can be readily automated, which makes it feasible to routinely synthesize peptides containing about 50 residues in good yield and purity. In fact, the solid-phase method has been used to synthesize interferons (155 residues) that have antiviral activity and ribonuclease (124 residues) that is catalytically active.

THREE-DIMENSIONAL PROTEIN STRUCTURE CAN BE DETERMINED BY NMR SPECTROSCOPY AND X-RAY CRYSTALLOGRAPHY

A crucial question is, What does the three-dimensional structure of a specific protein look like? Protein structure determines function, given that the specificity of active sites and binding sites depends on the precise threedimensional conformation. Nuclear magnetic resonance spectroscopy and x-ray crystallography are two of the most important techniques for elucidating the conformation of proteins.

NUCLEAR MAGNETIC RESONANCE SPECTROSCOPY CAN REVEAL THE STRUCTURES OF PROTEINS IN SOLUTION

Nuclear magnetic resonance (NMR) spectroscopy is unique in being able to reveal the *atomic structure* of macromolecules *in solution*, provided that highly concentrated solutions (~1 mM, or 15 mg ml^{-1} for a 15-kd protein) can be obtained. This technique depends on the fact that certain atomic nuclei are intrinsically magnetic. Only a limited number of isotopes display this property, called *spin*. The simplest example is the hydrogen nucleus (^1H), which is a proton. The spinning of a proton generates a magnetic moment. This moment can take either of two orientations, or spin states (called α and β), when an external magnetic field is applied. The energy difference between these states is proportional to the strength of the imposed magnetic field. The α state has a slightly lower energy and hence is slightly more populated (by a factor of the order of 1.00001 in a typical experiment) because it is aligned with the field. A spinning proton in an α state can be raised to an excited state (β state) by applying a pulse of electromagnetic radiation (a radio-frequency, or RF, pulse), provided the frequency corresponds to the energy difference between the α and the β states. In these circumstances, the spin will change from α to β; in other words, *resonance* will be obtained. A resonance spectrum for a molecule can be obtained by varying the magnetic field at a constant frequency of electromagnetic radiation or by keeping the magnetic field constant and varying electromagnetic radiation.

These properties can be used to examine the chemical surroundings of the hydrogen nucleus. The flow of electrons around a magnetic nucleus generates a small local magnetic field that opposes the applied field. The degree of such shielding depends on the surrounding electron density. Consequently, nuclei in different environments will change states, or resonate, at slightly different field strengths or radiation frequencies. The nuclei of the perturbed sample absorb electromagnetic radiation at a frequency that can be measured. The different frequencies, termed *chemical shifts,* are expressed in fractional units δ (parts per million, or ppm) relative to the shifts of a standard compound, such as a water-soluble derivative of tetramethysilane, that is added with the sample. For example, a -CH_3 proton typically exhibits a chemical shift Δ) of 1 ppm, compared with a chemical shift of 7 ppm for an aromatic proton. The chemical shifts of most protons in protein molecules fall between 0 and 9 ppm. It is possible to resolve most protons in many proteins by using this technique of *onedimensional NMR*. With this information, we can then deduce changes to a particular chemical group under different conditions, such as the conformational change of a protein from a disordered structure to an *α* helix in response to a change in pH.

We can garner even more information by examining how the spins on different protons affect their neighbours. By inducing a transient magnetization in a sample through the application a radio-frequency pulse, it is possible to alter the spin on one nucleus and examine the effect on the spin of a neighbouring nucleus. Especially revealing is a *two-dimensional spectrum obtained by nuclear Overhauser enhancement spectroscopy* (NOESY), *which graphically displays pairs of protons that are in close proximity,* even if they are not close together in the primary structure. The basis for this technique is the *nuclear Overhauser effect* (NOE), an interaction between nuclei that is proportional to the inverse sixth power of the distance between them. Magnetization is transferred from an excited nucleus to an unexcited one if they are less than about 5 Å apart. In other words, the effect provides a means of detecting the location of atoms relative to one another in the three-dimensional structure of the protein. The diagonal of a NOESY spectrum corresponds to a one-dimensional spectrum. The offdiagonal peaks provide crucial new information: *they identify pairs of protons that are less than 5 Å apart*. The large number of off-diagonal peaks reveals short proton-proton distances. The three-dimensional structure of a protein can be reconstructed with the use of such proximity relations. Structures are calculated such that protons that must be separated by less than 5 Å on the basis of NOESY spectra are close to one another in the three-dimensional structure. If a sufficient number of distance constraints are applied, the three-dimensional structure can be determined nearly uniquely. A family of related structures is generated for three reasons. First, not enough constraints may be experimentally accessible to fully specify the structure. Second, the distances obtained from analysis of the NOESY spectrum are only approximate. Finally, the experimental observations are made not on single molecules but on a large number of molecules in solution that may have slightly different structures at any given moment. Thus, the family of structures generated from NMR structure analysis indicates the range of conformations for the protein in solution. At present, NMR spectroscopy can determine the structures of only relatively small proteins (<40 kd), but its resolving power is certain to increase. The power of NMR has been greatly enhanced by the ability to produce proteins labeled uniformly or at specific sites with ^{13}C, ^{15}N, and ^{2}H with the use of recombinant DNA technology.

X-RAY CRYSTALLOGRAPHY REVEALS THREE-DIMENSIONAL STRUCTURE IN ATOMIC DETAIL

X-ray crystallography provides the finest visualization of protein structure currently available. This technique can reveal the precise three-dimensional positions of most atoms in a protein molecule. The use of x-rays provides the best resolution because the wavelength of x-rays is about the same length as that of a covalent bond. The three components in an x-ray crystallographic analysis are a *protein crystal*, a *source of x-rays*, and a *detector*.

The technique requires that all molecules be precisely oriented, so the first step is to obtain crystals of the protein of interest. Slowly adding ammonium sulfate or another salt to a concentrated solution of protein to reduce its solubility favours the formation of highly ordered crystals. For example, myoglobin crystallizes in 3 M ammonium sulfate. Some proteins crystallize readily, whereas others do so only after much effort has been expended in identifying the right conditions. Crystallization is an art; the best practitioners have great perseverance and patience. Increasingly large and complex proteins are being crystallized. For example, poliovirus, an 8500-kd assembly of 240 protein subunits surrounding an RNA core, has been crystallized and its structure solved by x-ray methods. Crucially, protein crystals frequently display their biological activity, indicating that the proteins have crystallized in their biologically active configuration. For instance, enzyme crystals may display catalytic activity if the crystals are suffused with substrate.

Next, a source of x-rays is required. A beam of x-rays of wavelength 1.54 Å is produced by accelerating electrons against a copper target. A narrow beam of x-rays strikes the protein crystal. Part of the beam goes straight through the crystal; the rest is *scattered* in various directions. Finally, these scattered, or *diffracted,* x-rays are detected by x-ray film, the blackening of the emulsion being proportional to the intensity of the scattered x-ray beam, or by a solid-state electronic detector. The scattering pattern provides abundant information about protein structure. The basic physical principles underlying the technique are:

1. *Electrons scatter x-rays.* The amplitude of the wave scattered by an atom is proportional to its number of electrons. Thus, a carbon atom scatters six times as strongly as a hydrogen atom does.
2. *The scattered waves recombine.* Each atom contributes to each scattered beam. The scattered waves reinforce one another at the film or detector if they are in phase (in step) there, and they cancel one another if they are out of phase.
3. *The way in which the scattered waves recombine depends only on the atomic arrangement.*

The protein crystal is mounted and positioned in a precise orientation with respect to the x-ray beam and the film. The crystal is rotated so that the beam can strike the crystal from many directions. This rotational motion results in an x-ray photograph consisting of a regular array of spots called *reflections*. The intensity of each spot is measured. These *intensities and their positions* are the basic experimental data of an x-ray crystallographic analysis. The next step is to reconstruct an image of the protein from the observed intensities. In light microscopy or electron microscopy, the diffracted beams are focused by lenses to directly form an image. However, appropriate lenses for focusing x-rays do not exist. Instead, the image is formed by applying a mathematical relation called a Fourier transform. For each spot, this operation yields a wave of electron density whose amplitude is proportional to the square root of the observed intensity of the spot. Each wave also has a *phase*—that is, the timing of its crests and troughs relative to those of other waves. The phase of each wave determines whether the wave reinforces or cancels the waves contributed by the other spots. These phases can be deduced from the well-understood diffraction patterns produced by electron-dense heavy-atom reference markers such as uranium or mercury at specific sites in the protein.

The stage is then set for the calculation of an electron-density map, which gives the density of electrons at a large number of regularly spaced points in the crystal. This three-dimensional electron-density distribution is represented by a series of parallel s stacked on top of one another. Each is a transparent plastic sheet (or, more recently, a layer in a computer image) on which the electron-density distribution is represented by contour lines, like the contour lines used in geological survey maps to depict altitude. The next step is to interpret the electron-density map. A critical factor is the *resolution* of the x-ray analysis, which is determined by the number of scattered intensities used in the Fourier synthesis. The fidelity of the image depends on the resolution of the Fourier synthesis, as shown by the optical analogy. A resolution of 6 Å reveals the course of the polypeptide chain but few other structural details. The reason is that polypeptide chains pack together so that their centres are between 5 Å and 10 Å apart. Maps at higher resolution are needed to delineate groups of atoms, which lie between 2.8 Å and 4.0 Å apart, and individual atoms, which are between 1.0 Å and 1.5 Å apart. The ultimate resolution of an x-ray analysis is determined by the degree of perfection of the crystal. For proteins, this limiting resolution is usually about 2 Å.

The structures of more than 10,000 proteins had been elucidated by NMR and x-ray crystallography by mid-2000, and several new structures are now determined each day. The coordinates are collected at the Protein Data Bank and the structures can be accessed for visualization and analysis. Knowledge of the detailed molecular architecture of proteins has been a source of insight into how proteins recognize and bind other molecules, how they function as enzymes, how they fold, and how they evolved. This extraordinarily rich harvest is continuing at a rapid pace and is greatly influencing the entire field of biochemistry.

Chapter 6

DNA, RNA, and the Flow of Genetic Information

DNA and RNA are long linear polymers, called nucleic acids, that carry information in a form that can be passed from one generation to the next. These macromolecules consist of a large number of linked nucleotides, each composed of a sugar, a phosphate, and a base. Sugars linked by phosphates form a common backbone, whereas the bases vary among four kinds. *Genetic information is stored in the sequence of bases along a nucleic acid chain.* The bases have an additional special property: they form specific pairs with one another that are stabilized by hydrogen bonds. The base pairing results in the formation of a double helix, a helical structure consisting of two strands. *These base pairs provide a mechanism for copying the genetic information in an existing nucleic acid chain to form a new chain.* Although RNA probably functioned as the genetic material very early in evolutionary history, the genes of all modern cells and many viruses are made of DNA. DNA is replicated by the action of DNA polymerase enzymes. These exquisitely specific enzymes copy sequences from nucleic acid templates with an error rate of less than 1 in 100 million nucleotides.

Genes specify the kinds of proteins that are made by cells, but DNA is not the direct template for protein synthesis. Rather, the templates for protein synthesis are RNA (ribonucleic acid) molecules. In particular, a class of RNA molecules called *messenger RNA* (mRNA) are the information-carrying intermediates in protein synthesis. Other RNA molecules, such as *transfer RNA* (tRNA) and *ribosomal RNA* (rRNA), are part of the protein-synthesizing machinery. All forms of cellular RNA are synthesized by RNA polymerases that take instructions from DNA templates. This process of *transcription* is followed by *translation*, the synthesis of proteins according to instructions given by mRNA templates. Thus, the flow of genetic information, or *gene expression*, in normal cells is:

This flow of information is dependent on the genetic code, which defines the relation between the sequence of bases in DNA (or its mRNA transcript) and the sequence of amino acids in a protein. The code is nearly the same in all organisms: a sequence of three bases, called a *codon*, specifies an amino acid. Codons in mRNA are read sequentially by tRNA molecules, which serve as adaptors in protein synthesis. Protein synthesis takes place on ribosomes, which are complex assemblies of rRNAs and more than 50 kinds of proteins.

The last theme to be considered is the interrupted character of most eukaryotic genes, which are mosaics of nucleic acid sequences called *introns* and *exons*. Both are transcribed, but introns are cut out of newly synthesized RNA molecules, leaving mature RNA molecules with continuous exons. The existence of introns and exons has crucial implications for the evolution of proteins.

A NUCLEIC ACID CONSISTS OF FOUR KINDS OF BASES LINKED TO A SUGAR-PHOSPHATE BACKBONE

The nucleic acids DNA and RNA are well suited to function as the carriers of genetic information by virtue of their covalent structures. These macromolecules are *linear polymers* built up from similar units connected end to end. Each monomer unit within the polymer consists of three components: a sugar, a phosphate, and a base. The sequence of bases uniquely characterizes a nucleic acid and represents a form of linear information.

RNA AND DNA DIFFER IN THE SUGAR COMPONENT AND ONE OF THE BASES

The sugar in *deoxyribonucleic acid (DNA)* is *deoxyribose*. The deoxy prefix indicates that the 2′ carbon atom of the sugar lacks the oxygen atom that is linked to the 2′ carbon atom of *ribose* (the sugar in *ribonucleic acid,* or *RNA*). The sugars in nucleic acids are linked to one another by phosphodiester bridges. Specifically, the 3′ -hydroxyl (3′ -OH) group of the sugar moiety of one nucleotide is esterified to a phosphate group, which is, in turn, joined to the 5′ -hydroxyl group of the adjacent sugar. The chain of sugars linked by phosphodiester bridges is referred to as the *backbone* of the nucleic acid. Whereas the backbone is constant in DNA and RNA, the bases vary from one monomer to the next. Two of the bases are derivatives of *purine*—adenine (A) and guanine (G)—and two of *pyrimidine*—cytosine (C) and thymine (T, DNA only) or uracil (U, RNA only).

RNA, like DNA, is a long unbranched polymer consisting of nucleotides joined by 3′→ 5′ phosphodiester bonds. The covalent structure of RNA differs from that of DNA in two respects. As stated earlier and as indicated by its name, the sugar units in RNA are riboses rather than deoxyriboses. Ribose contains a 2′ -hydroxyl group not present in deoxyribose. As a consequence, in addition to the standard 3′ →5′ linkage, a 2′ →5′ linkage is possible for RNA. This later linkage is important in the removal of introns and the joining of exons for the formation of mature RNA. The other difference, as already mentioned, is that one of the four major bases in RNA is uracil (U) instead of thymine (T).

Note that each phosphodiester bridge has a negative charge. This negative charge repels nucleophilic species such as hydroxide ion; consequently, phosphodiester linkages are much less susceptible to hydrolytic attack than are other esters such as carboxylic acid esters. This resistance is crucial for maintaining the integrity of information stored in nucleic acids. The absence of the 2′ -hydroxyl group in DNA further increases its resistance to hydrolysis. The greater stability of

DNA probably accounts for its use rather than RNA as the hereditary material in all modern cells and in many viruses.

NUCLEOTIDES ARE THE MONOMERIC UNITS OF NUCLEIC ACIDS

A unit consisting of a base bonded to a sugar is referred to as a *nucleoside*. The four nucleoside units in RNA are called *adenosine, guanosine, cytidine,* and *uridine,* whereas those in DNA are called *deoxyadenosine, deoxyguanosine, deoxycytidine,* and *thymidine.* In each case, N-9 of a purine or N-1 of a pyrimidine is attached to C-1′ of the sugar. The base lies above the plane of sugar when the structure is written in the standard orientation; that is, the configuration of the *N*-glycosidic linkage is *β*. A *nucleotide* is a nucleoside joined to one or more phosphate groups by an ester linkage. The most common site of esterification in naturally occurring nucleotides is the hydroxyl group attached to C-5′ of the sugar. A compound formed by the attachment of a phosphate group to the C-5′ of a nucleoside sugar is called a *nucleoside 5′ -phosphate* or a *5′ -nucleotide*. For example, ATP is *adenosine 5′ -triphosphate*. Another nucleotide is deoxyguanosine 3′ -monophosphate. This nucleotide differs from ATP in that it contains guanine rather than adenine, contains deoxyribose rather than ribose (indicated by the prefix "d"), contains one rather than three phosphates, and has the phosphate esterified to the hydroxyl group in the 3′ rather than the 5′ position. Nucleotides are the monomers that are linked to form RNA and DNA. The four nucleotide units in DNA are called *deoxyadenylate, deoxyguanylate, deoxycytidylate,* and *deoxythymidylate,* and *thymidylate.* Note that thymidylate contains deoxyribose; by convention, the prefix deoxy is not added because thymine-containing nucleotides are only rarely found in RNA.

The abbreviated notations pApCpG or pACG denote a trinucleotide of DNA consisting of the building blocks deoxyadenylate monophosphate, deoxycytidylate monophosphate, and deoxyguanylate monophosphate linked by a phosphodiester bridge, where "p" denotes a phosphate group. The 5′ end will often have a phosphate attached to the 5′ -OH group. Note that, like a polypeptide *a DNA chain has polarity*. One end of the chain has a free 5′ -OH group (or a 5′ -OH group attached to a phosphate), whereas the other end has a 3′ -OH group, neither of which is linked to another nucleotide. By convention, *the base sequence is written in the 5′ -to-32 direction*. Thus, the symbol ACG indicates that the unlinked 5′ -OH group is on deoxyadenylate, whereas the unlinked 3′ -OH group is on deoxyguanylate. Because of this polarity, ACG and GCA correspond to different compounds.

A striking characteristic of naturally occurring DNA molecules is their length. A DNA molecule must comprise many nucleotides to carry the genetic information necessary for even the simplest organisms. For example, the DNA of a virus such as polyoma, which can cause cancer in certain organisms, is as long as 5100 nucleotides in length. We can quantify the information carrying capacity of nucleic acids in the following way. Each position can be one of four bases, corresponding to two bits of information ($2^2 = 4$). Thus, a chain of 5100 nucleotides corresponds to $2 \times 5100 = 10{,}200$ bits, or 1275 bytes (1 byte = 8 bits). The *E. coli* genome is a single DNA molecule consisting of two chains of 4.6 million nucleotides, corresponding to 9.2 million bits, or 1.15 megabytes, of information.

DNA molecules from higher organisms can be much larger. The human genome comprises approximately 3 billion nucleotides, divided among 24 distinct DNA molecules (22 autosomes, x and y sex chromosomes) of different sizes. One of the largest known DNA molecules is found in the Indian muntjak, an Asiatic deer; its genome is nearly as large as the human genome but is distributed on only 3 chromosomes. The largest of these chromosomes has chains of more than 1 billion nucleotides. If such a DNA molecule could be fully extended, it would stretch more than 1 foot in length. Some plants contain even larger DNA molecules.

A PAIR OF NUCLEIC ACID CHAINS WITH COMPLEMENTARY SEQUENCES CAN FORM A DOUBLE-HELICAL STRUCTURE

The covalent structure of nucleic acids accounts for their ability to carry information in the form of a sequence of bases along a nucleic acid chain. Other features of nucleic acid structure facilitate the process of *replication*—that is, the generation of two copies of a nucleic acid from one. These features depend on the ability of the bases found in nucleic acids to form *spe-cific base pairs* in such a way that a helical structure consisting of two strands is formed. The double-helical structure of DNA facilitates the replication of the genetic material.

THE DOUBLE HELIX IS STABILIZED BY HYDROGEN BONDS AND HYDROPHOBIC INTERACTIONS

The existence of specific base-pairing interactions was discovered in the course of studies directed at determining the three-dimensional structure of DNA. Maurice Wilkins and Rosalind Franklin obtained x-ray diffraction photographs of fibers of DNA. The characteristics of these diffraction patterns indicated that DNA was formed of two chains that wound in a regular helical structure. From these and other data, James Watson and Francis Crick inferred a structural model for DNA that accounted for the diffraction pattern and was also the source of some remarkable insights into the functional properties of nucleic acids.

The features of the Watson-Crick model of DNA deduced from the diffraction patterns are:

1. Two helical polynucleotide chains are coiled around a common axis. The chains run in opposite directions.
2. The sugar-phosphate backbones are on the outside and, therefore, the purine and pyrimidine bases lie on the inside of the helix.
3. The bases are nearly perpendicular to the helix axis, and adjacent bases are separated by 3.4 Å. The helical

structure repeats every 34 Å, so there are 10 bases (= 34 Å per repeat/3.4 Å per base) per turn of helix. There is a rotation of 36 degrees per base (360 degrees per full turn/10 bases per turn).

4. The diameter of the helix is 20 Å.

How is such a regular structure able to accommodate an arbitrary sequence of bases, given the different sizes and shapes of the purines and pyrimidines? In attempting to answer this question, Watson and Crick discovered that guanine can be paired with cytosine and adenine with thymine to form base pairs that have essentially the same shape. These base pairs are held together by specific hydrogen bonds. This base-pairing scheme was supported by earlier studies of the base composition of DNA from different species. In 1950, Erwin Chargaff reported that the ratios of adenine to thymine and of guanine to cytosine were nearly the same in all species studied. Note that all the adenine:thymine and guanine:cytosine ratios are close to 1, whereas the adenine-to-guanine ratio varies considerably. The meaning of these equivalences was not evident until the Watson-Crick model was proposed, when it became clear that they represent an essential facet of DNA structure.

The spacing of approximately 3.4 Å between nearly parallel base pairs is readily apparent in the DNA diffraction pattern. The stacking of bases one on top of another contributes to the stability of the double helix in two ways. First, adjacent base pairs attract one another through van der Waals forces (.3.). Energies associated with van der Waals interactions are quite small, such that typical interactions contribute from 0.5 to 1.0 kcal mol^{-1} per atom pair. In the double helix, however, a large number of atoms are in van der Waals contact, and the net effect, summed over these atom pairs, is substantial. In addition, the double helix is stabilized by the hydrophobic effect (.3.4): base stacking, or hydrophobic interactions between the bases, results in the exposure of the more polar surfaces to the surrounding water. This arrangement is reminiscent of protein folding, where hydrophobic amino acids are interior in the protein and hydrophilic are exterior (3.4). Base stacking in DNA is also favoured by the conformations of the relatively rigid five-membered rings of the backbone sugars. The sugar rigidity affects both the single-stranded and the double-helical forms.

THE DOUBLE HELIX FACILITATES THE ACCURATE TRANSMISSION OF HEREDITARY INFORMATION

The double-helical model of DNA and the presence of specific base pairs immediately suggested how the genetic material might replicate. The sequence of bases of one strand of the double helix precisely determines the sequence of the other strand; a guanine base on one strand is always paired with a cytosine base on the other strand, and so on. Thus, separation of a double helix into its two component chains would yield two single-stranded templates onto which new double helices could be constructed, each of which would have the same sequence of bases as the parent double helix. Consequently, as DNA is replicated, one of the chains of each daughter DNA molecule would be newly synthesized, whereas the other would be passed unchanged from the parent DNA molecule. This distribution of parental atoms is achieved by *semiconservative replication*..

Matthew Meselson and Franklin Stahl carried out a critical test of this hypothesis in 1958. They labeled the parent DNA with ^{15}N, a heavy isotope of nitrogen, to make it denser than ordinary DNA. The labeled DNA was generated by growing *E. coli* for many generations in a medium that contained $^{15}NH_4Cl$ as the sole nitrogen source. After the incorporation of heavy nitrogen was complete, the bacteria were abruptly transferred to a medium that contained ^{14}N, the ordinary isotope of nitrogen. The question asked was: What is the distribution of ^{14}N and ^{15}N in the DNA molecules after successive rounds of replication?

The distribution of ^{14}N and ^{15}N was revealed by the technique of *density-gradient equilibrium sedimentation*. A small amount of DNA was dissolved in a concentrated solution of cesium chloride having a density close to that of the DNA (1.7 g cm^{-3}). This solution was centrifuged until it was nearly at equilibrium. The opposing processes of sedimentation and diffusion created a gradient in the concentration of cesium chloride across the centrifuge cell. The result was a stable density gradient, ranging from 1.66 to 1.76 g cm^{-3}. The DNA molecules in this density gradient were driven by centrifugal force into the region where the solution's density was equal to their own. The genomic DNA yielded a narrow band that was detected by its absorption of ultraviolet light. A mixture of ^{14}N DNA and ^{15}N DNA molecules gave clearly separate bands because they differ in density by about 1per cent (5.4).

DNA was extracted from the bacteria at various times after they were transferred from a ^{15}N to a ^{14}N medium and centrifuged. Analysis of these samples showed that there was a single band of DNA after one generation. The density of this band was precisely halfway between the densities of the ^{14}N DNA and ^{15}N DNA bands. The absence of ^{15}N DNA indicated that parental DNA was not preserved as an intact unit after replication. The absence of ^{14}N DNA indicated that all the daughter DNA derived some of their atoms from the parent DNA. This proportion had to be half because the density of the hybrid DNA band was halfway between the densities of the ^{14}N DNA and ^{15}N DNA bands.

After two generations, there were equal amounts of two bands of DNA. One was hybrid DNA, and the other was ^{14}N DNA. Meselson and Stahl concluded from these incisive experiments "that the nitrogen in a DNA molecule is divided equally between two physically continuous subunits; that following duplication, each daughter molecule receives one of these; and that the subunits are conserved through many duplications." Their results agreed perfectly with the Watson-Crick model for DNA replication.

THE DOUBLE HELIX CAN BE REVERSIBLY MELTED

During DNA replication and other processes, the two strands of the double helix must be separated from one another, at least in a local region. In the laboratory, the double helix can be disrupted by heating a solution of DNA. The heating disrupts the hydrogen bonds between base pairs and thereby causes the strands to separate. The dissociation of the double helix is often called *melting* because it occurs relatively abruptly at a certain temperature. The *melting temperature* (T_m) is defined as the temperature at which half the helical structure is lost. Strands may also be separated by adding acid or alkali to ionize the nucleotide bases and disrupt base pairing.

Stacked bases in nucleic acids absorb less ultraviolet light than do unstacked bases, an effect called *hypochromism*. Thus, the melting of nucleic acids is easily followed by monitoring their absorption of light, which peaks at a wavelength of 260 nm.

Separated complementary strands of nucleic acids spontaneously reassociate to form a double helix when the temperature is lowered below T_m. This renaturation process is sometimes called *annealing*. The facility with which double helices can be melted and then reassociated is crucial for the biological functions of nucleic acids. Of course, inside cells, the double helix is not melted by the addition of heat. Instead, proteins called *helicases* use chemical energy (from ATP) to disrupt the structure of double-stranded nucleic acid molecules.

The ability to reversibility melt and reanneal DNA in the laboratory provides a powerful tool for investigating sequence similarity as well as gene structure and expression. For instance, DNA molecules from two different organisms can be melted and allowed to reanneal or *hybridize* in the presence of each other. If the sequences are similar, hybrid DNA duplexes, with DNA from each organism contributing a strand of the double helix, can form. Indeed, the degree of hybridization is an indication of the relatedness of the genomes and hence the organisms. Similar hybridization experiments with RNA and DNA can locate genes in a cell's DNA that correspond to a particular RNA. We will return to this important technique.

SOME DNA MOLECULES ARE CIRCULAR AND SUPERCOILED

The DNA molecules in human chromosomes are linear. However, electron microscopic and other studies have shown that intact DNA molecules from some other organisms are circular. The term *circular* refers to the continuity of the DNA chains, not to their geometrical form. DNA molecules inside cells necessarily have a very compact shape. Note that the *E. coli* chromosome, fully extended, would be about 1000 times as long as the greatest diameter of the bacterium.

A new property appears in the conversion of a linear DNA molecule into a closed circular molecule. The axis of the double helix can itself be twisted into a *superhelix*. A circular DNA molecule without any superhelical turns is known as a *relaxed molecule*. Supercoiling is biologically important for two reasons. First, *a supercoiled DNA molecule has a more compact shape than does its relaxed counterpart*. Second, *supercoiling may hinder or favour the capacity of the double helix to unwind and thereby affects the interactions between DNA and other molecules*. These topological features of DNA will be considered further in 27.3.

SINGLE-STRANDED NUCLEIC ACIDS CAN ADOPT ELABORATE STRUCTURES

Single-stranded nucleic acids often fold back on themselves to form well-defined structures. Early in evolutionary history, nucleic acids, particularly RNA, may have adopted complex and diverse structures both to store genetic information and to catalyze its transmission. Such structures are also important in all modern organisms in entities such as the ribosome, a large complex of RNAs and proteins on which proteins are synthesized.

The simplest and most common structural motif formed is a *stem-loop*, created when two complementary sequences within a single strand come together to form double-helical structures. In many cases, these double helices are made up entirely of Watson-Crick base pairs. In other cases, however, the structures include mismatched or unmatched (bulged) bases. Such mismatches destabilize the local structure but introduce deviations from the standard double-helical structure that can be important for higher-order folding and for function.

Single-stranded nucleic acids can adopt structures more complex than simple stem-loops through the interaction of more widely separated bases. Often, three or more bases may interact to stabilize these structures. In such cases, hydrogen-bond donors and acceptors that ordinarily participate in Watson-Crick base pairs may participate in hydrogen bonds of nonstandard pairings. Metal ions such as magnesium ion (Mg^{2+}) often assist in the stabilization of these more elaborate structures.

DNA IS REPLICATED BY POLYMERASES THAT TAKE INSTRUCTIONS FROM TEMPLATES

We now turn to the molecular mechanism of DNA replication. The full replication machinery in cells comprises more than 20 proteins engaged in intricate and coordinated interplay. In 1958, Arthur Kornberg and his colleagues isolated the first known of the enzymes, called DNA polymerases, that promote the formation of the bonds joining units of the DNA backbone.

DNA POLYMERASE CATALYZES PHOSPHODIESTER-BOND FORMATION

DNA polymerases catalyze the step-by-step addition of deoxyribonucleotide units to a DNA chain. Importantly, *the new DNA chain is assembled directly on a preexisting DNA template*. The reaction catalyzed, in its simplest form, is:

Where dNTP stands for any deoxyribonucleotide and PP_i is a pyrophosphate molecule. The template can be a single strand of DNA or a double strand with one of the chains broken at one or more sites. If single stranded, the template DNA must be bound to a *primer* strand having a free 3′-hydroxyl group. The reaction also requires all four activated precursors—that is, the deoxynucleoside 5′-triphosphates dATP, dGTP, dTTP, and dCTP—as well as Mg^{2+} ion.

The chain-elongation reaction catalyzed by DNA polymerases is a nucleophilic attack by the 32 -hydroxyl group of the primer on the innermost phosphorus atom of the deoxynucleoside triphosphate (5.22). A phosphodiester bridge forms with the concomitant release of pyrophosphate. The subsequent hydrolysis of pyrophosphate by pyrophosphatase, a ubiquitous enzyme, helps drive the polymerization forward. Elongation of the DNA chain proceeds in the 5′-to-3′ direction.

DNA polymerases catalyze the formation of a phosphodiester bond efficiently only if the base on the incoming nucleoside triphosphate is complementary to the base on the template strand. Thus, DNA polymerase is a *template-directed enzyme* that synthesizes a product with a base sequence complementary to that of the template. Many DNA polymerases also have a separate nuclease activity that allows them to correct mistakes in DNA by using a different reaction to remove mismatched nucleotides. These properties of DNA polymerases contribute to the remarkably high fidelity of DNA replication, which has an error rate of less than 10^{-8} per base pair.

THE GENES OF SOME VIRUSES ARE MADE OF RNA

Genes in all cellular organisms are made of DNA. The same is true for some viruses, but for others the genetic material is RNA. Viruses are genetic elements enclosed in protein coats that can move from one cell to another but are not capable of independent growth. One well-studied example of an RNA virus is the tobacco mosaic virus, which infects the leaves of tobacco plants. This virus consists of a single strand of RNA (6930 nucleotides) surrounded by a protein coat of 2130 identical subunits. An RNA-directed RNA polymerase catalyzes the replication of this viral RNA.

Another important class of RNA virus comprises the *retroviruses,* so called because the genetic information flows from RNA to DNA rather than from DNA to RNA. This class includes human immunodeficiency virus 1 (HIV-1), the cause of AIDS, as well as a number of RNA viruses that produce tumors in susceptible animals. Retrovirus particles contain two copies of a single-stranded RNA molecule. On entering the cell, the RNA is copied into DNA through the action of a viral enzyme called *reverse transcriptase*. The resulting double-helical DNA version of the viral genome can become incorporated into the chromosomal DNA of the host and is replicated along with the normal cellular DNA. At a later time, the integrated viral genome is expressed to form viral RNA and viral proteins, which assemble into new virus particles.

Note that RNA viruses are not vestiges of the RNA world. Instead, fragments of RNA in these viruses have evolved to encode their protein coats and other structures needed for transferring from cell to cell and replicating.

GENE EXPRESSION IS THE TRANSFORMATION OF DNA INFORMATION INTO FUNCTIONAL MOLECULES

The information stored as DNA becomes useful when it is expressed in the production of RNA and proteins. This rich and complex topic is the subject of several chapters later in this book, but here we introduce the basics of gene expression. DNA can be thought of as archival information, stored and manipulated judiciously to minimize damage (mutations). It is expressed in two steps. First, an RNA copy is made. An RNA molecule that encodes proteins can be thought of as a photocopy of the original information—it can be made in multiple copies, used, and then disposed of. Second, an RNA molecule can be further thought of as encoding directions for protein synthesis that must be translated to be of use. The information in messenger RNA is translated into a functional protein. Other types of RNA molecules exist to facilitate this translation. We now examine the transcription of DNA information into RNA, the translation of RNA information into protein, and the genetic code that links nucleotide sequence with amino acid sequence.

SEVERAL KINDS OF RNA PLAY KEY ROLES IN GENE EXPRESSION

1. *Messenger RNA* is the template for protein synthesis or *translation*. An mRNA molecule may be produced for each gene or group of genes that is to be expressed in *E. coli,* whereas a distinct mRNA is produced for each gene in eukaryotes. Consequently, mRNA is a heterogeneous class of molecules. In *E. coli,* the average length of an mRNA molecule is about 1.2 kilobases (kb).
2. *Transfer RNA* carries amino acids in an activated form to the ribosome for peptide-bond formation, in a sequence dictated by the mRNA template. There is at least one kind of tRNA for each of the 20 amino acids. Transfer RNA consists of about 75 nucleotides (having a mass of about 25 kd), which makes it the smallest of the RNA molecules.
3. *Ribosomal RNA (rRNA),*the major component of ribosomes, plays both a catalytic and a structural role in protein synthesis (29.3.1). In *E. coli,* there are three kinds of rRNA, called *23S, 16S,* and *5S RNA* because of their sedimentation behaviour. One molecule of each of these species of rRNA is present in each ribosome.

Ribosomal RNA is the most abundant of the three types of RNA. Transfer RNA comes next, followed by messenger RNA, which constitutes only 5per cent of the total RNA. Eukaryotic cells contain additional small RNA molecules. *Small nuclear RNA* (snRNA) molecules, for example, participate in the splicing of RNA exons. A small RNA molecule in the cytosol plays a role in the targeting of newly synthesized proteins to intracellular compartments and extracellular destinations.

All Cellular RNA Is Synthesized by RNA Polymerases.

The synthesis of RNA from a DNA template is called *transcription* and is catalyzed by the enzyme *RNA polymerase.* RNA polymerase requires the following components:

1. *A template.* The preferred template is *double-stranded DNA*. Single-stranded DNA also can serve as a template. RNA, whether single or double stranded, is not an effective template; nor are RNA-DNA hybrids.
2. *Activated precursors.* All four *ribonucleoside triphosphates*—ATP, GTP, UTP, and CTP—are required.
3. *A divalent metal ion.* Mg^{2+} or Mn^{2+} are effective.

RNA polymerase catalyzes the initiation and elongation of RNA chains. The reaction catalyzed by this enzyme is:

The synthesis of RNA is like that of DNA in several respects. First, the direction of synthesis is $5' \rightarrow 3'$. Second, the mechanism of elongation is similar: the 3′ – OH group at the terminus of the growing chain makes a nucleophilic attack on the innermost phosphate of the incoming nucleoside triphosphate. Third, the synthesis is driven forward by the hydrolysis of pyrophosphate. In contrast with DNA polymerase, however, RNA polymerase does not require a primer. In addition, RNA polymerase lacks the nuclease capability used by DNA polymerase to excise mismatched nucleotides.

All three types of cellular RNA—mRNA, tRNA, and rRNA—are synthesized in *E. coli* by the same RNA polymerase according to instructions given by a DNA template. In mammalian cells, there is a division of labour among several different kinds of RNA polymerases.

RNA POLYMERASES TAKE INSTRUCTIONS FROM DNA TEMPLATES

RNA polymerase, like the DNA polymerases described earlier, takes instructions from a DNA template. The earliest evidence was the finding that the *base composition* of newly synthesized RNA is the complement of that of the DNA template strand, as exemplified by the RNA synthesized from a template of single-stranded φ × 174 DNA. *Hybridization experiments* also revealed that RNA synthesized by RNA polymerase is complementary to its DNA template. In these experiments, DNA is melted and allowed to reassociate in the presence of mRNA. RNA-DNA hybrids will form if the RNA and DNA have complementary sequences. The strongest evidence for the fidelity of transcription came from base-sequence studies showing that the RNA sequence is the precise complement of the DNA template sequence.

TRANSCRIPTION BEGINS NEAR PROMOTER SITES AND ENDS AT TERMINATOR SITES

RNA polymerase must detect and transcribe discrete genes from within large stretches of DNA. What marks the beginning of a transcriptional unit? DNA templates contain regions called *promoter sites* that specifically bind RNA polymerase and determine where transcription begins. In bacteria, two sequences on the 5′ (upstream) side of the first nucleotide to be transcribed function as promoter sites. One of them, called the *Pribnow box,* has the consensus sequence TATAAT and is centred at –10 (10 nucleotides on the 5′ side of the first nucleotide transcribed, which is denoted by + 1). The other, called the *-35 region,* has the consensus sequence TTGACA. The first nucleotide transcribed is usually a purine.

Eukaryotic genes encoding proteins have promoter sites with a TATAAA consensus sequence, called a *TATA box* or a *Hogness box,* centred at about –25. Many eukaryotic promoters also have a *CAAT box* with a GGN*CAAT*CT consensus sequence centred at about -75. Transcription of eukaryotic genes is further stimulated by *enhancer sequences,* which can be quite distant (as many as several kilobases) from the start site, on either its 5′ or its 3′ side.

RNA polymerase proceeds along the DNA template, transcribing one of its strands until it reaches a terminator sequence. This sequence encodes a termination signal, which in *E. coli* is a *base-paired hairpin* on the newly synthesized RNA molecule. This hairpin is formed by base pairing of self-complementary sequences that are rich in G and C. Nascent

RNA spontaneously dissociates from RNA polymerase when this hairpin is followed by a string of U residues. Alternatively, RNA synthesis can be terminated by the action of *rho,* a protein. Less is known about the termination of transcription in eukaryotes. The important point now is that *discrete start and stop signals for transcription are encoded in the DNA template.*

In eukaryotes, the mRNA is modified after transcription. A "cap" structure is attached to the 5′ end, and a sequence of adenylates the poly(A) tail is added to the 3′ end.

TRANSFER RNA IS THE ADAPTOR MOLECULE IN PROTEIN SYNTHESIS

We have seen that mRNA is the template for protein synthesis. How then does it direct amino acids to become joined in the correct sequence to form a protein? In 1958, Francis Crick wrote:

> RNA presents mainly a sequence of sites where hydrogen bonding could occur. One would expect, therefore, that whatever went onto the template in a *specific* way did so by forming hydrogen bonds. It is therefore a natural hypothesis that the amino acid is carried to the template by an adaptor molecule, and that the adaptor is the part that actually fits onto the RNA. In its simplest form, one would require twenty adaptors, one for each amino acid.

This highly innovative hypothesis soon became established as fact. *The adaptor in protein synthesis is transfer RNA.* For the moment, it suffices to note that tRNA contains an *amino acidattachment site* and a *template-recognition site.* A tRNA molecule carries a specific amino acid in an activated form to the site of protein synthesis. The carboxyl group of this amino acid is esterified to the 3′ - or 2′ -hydroxyl group of the ribose unit at the 3′ end of the tRNA chain. The joining of an amino acid to a tRNA molecule to form an *aminoacyl-tRNA* is catalyzed by a specific enzyme called an *aminoacyl-tRNA synthetase* (or *acti-vating enzyme*). This esterification reaction is driven by ATP. There is at least one specific synthetase for each of the 20 amino acids. The template-recognition site on tRNA is a sequence of three bases called an *anticodon*. The anticodon on tRNA recognizes a complementary sequence of three bases, called a *codon*, on mRNA.

AMINO ACIDS ARE ENCODED BY GROUPS OF THREE BASES STARTING FROM A FIXED POINT

The *genetic code* is the relation between the sequence of bases in DNA (or its RNA transcripts) and the sequence of amino acids in proteins. Experiments by Francis Crick, Sydney Brenner, and others established the following features of the genetic code by 1961:

1. *Three nucleotides encode an amino acid*. Proteins are built from a basic set of 20 amino acids, but there are only four bases. Simple calculations show that a minimum of three bases is required to encode at least 20 amino acids. Genetic experiments showed that *an amino acid is in fact encoded by a group of three bases, or codon*.
2. *The code is nonoverlapping*. Consider a base sequence ABCDEF. In an overlapping code, ABC specifies the first amino acid, BCD the next, CDE the next, and so on. In a nonoverlapping code, ABC designates the first amino acid, DEF the second, and so forth. Genetics experiments again established the code to be nonoverlapping.
3. *The code has no punctuation*. In principle, one base (denoted as Q) might serve as a "comma" between groups of three bases.

This is not the case. Rather, *the sequence of bases is read sequentially from a fixed starting point*, without punctuation.

4. *The genetic code is degenerate*. Some amino acids are encoded by more than one codon, inasmuch as there are 64 possible base triplets and only 20 amino acids. In fact, 61 of the 64 possible triplets specify particular amino acids and 3 triplets (called stop codons)

designate the termination of translation. Thus, *for most amino acids, there is more than one code word.*

5.5.1. MAJOR FEATURES OF THE GENETIC CODE

All 64 codons have been deciphered. Because the code is highly degenerate, only tryptophan and methionine are encoded by just one triplet each. The other 18 amino acids are each encoded by two or more. Indeed, leucine, arginine, and serine are specified by six codons each. The number of codons for a particular amino acid correlates with its frequency of occurrence in proteins.

Codons that specify the same amino acid are called *synonyms*. For example, CAU and CAC are synonyms for histidine. Note that synonyms are not distributed haphazardly throughout the genetic code. An amino acid specified by two or more synonyms occupies a single box (unless it is specified by more than four synonyms). The amino acids in a box are specified by codons that have the same first two bases but differ in the third base, as exemplified by GUU, GUC, GUA, and GUG. Thus, *most synonyms differ only in the last base of the triplet*. Inspection of the code shows that XYC and XYU always encode the same amino acid, whereas XYG and XYA usually encode the same amino acid. The structural basis for these equivalences of codons will become evident when we consider the nature of the anticodons of tRNA molecules.

What is the biological significance of the extensive degeneracy of the genetic code? If the code were not degenerate, 20 codons would designate amino acids and 44 would lead to chain termination. The probability of mutating to chain termination would therefore be much higher with a nondegenerate code. Chain-termination mutations usually lead to inactive proteins, whereas substitutions of one amino acid for another are usually rather harmless. Thus, *degeneracy minimizes the deleterious effects of mutations*. Degeneracy of the code may also be significant in permitting DNA base composition to vary over a wide range without altering the amino acid sequence of the proteins encoded by the DNA. The G + C content of bacterial DNA ranges from less than 30per cent to more than 70per cent. DNA molecules with quite different G + C contents could encode the same proteins if different synonyms of the genetic code were consistently used.

MESSENGER RNA CONTAINS START AND STOP SIGNALS FOR PROTEIN SYNTHESIS

Messenger RNA is translated into proteins on *ribosomes,* large molecular complexes assembled from proteins and ribosomal RNA. How is mRNA interpreted by the translation apparatus? As already mentioned, *UAA, UAG, and UGA designate chain termination.* These codons are read not by tRNA molecules but rather by specific proteins called *release factors.* Binding of the release factors to the ribosomes releases the newly synthesized protein. The start signal for protein synthesis is more complex. Polypeptide chains in bacteria start with a modified amino acid—namely, formylmethionine (fMet). A specific tRNA, the initiator tRNA, carries fMet. This fMet-tRNA recognizes the codon AUG or, less frequently, GUG. However, AUG is also the codon for an internal methionine residue, and GUG is the codon for an internal valine residue. Hence, the signal for the first amino acid in a prokaryotic polypeptide chain must be more complex than that for all subsequent ones. *AUG (or GUG) is only part of the initiation signal.* In bacteria, the initiating AUG (or GUG) codon is preceded several nucleotides away by a purine-rich sequence that base-pairs with a complementary sequence in a ribosomal RNA molecule. In eukaryotes, the AUG closest to the 5′ end of an mRNA molecule is usually the start signal for protein synthesis. This particular AUG is read by an initiator tRNA conjugated to methionine. Once the initiator AUG is located, the *reading frame* is established—groups of three nonoverlapping nucleotides are defined, beginning with the initiator AUG codon.

THE GENETIC CODE IS NEARLY UNIVERSAL

Is the genetic code the same in all organisms? The base sequences of many wild-type and mutant genes are known, as are the amino acid sequences of their encoded proteins. In each case, the nucleotide change in the gene and the amino acid change in the protein are as predicted by the genetic code. Furthermore, mRNAs can be correctly translated by the proteinsynthesizing machinery of very different species. For example, human hemoglobin mRNA is correctly translated by a wheat germ extract, and bacteria efficiently express recombinant DNA molecules encoding human proteins such as insulin. These experimental findings strongly suggested that the genetic code is universal.

A surprise was encountered when the sequence of human mitochondrial DNA became known. Human mitochondria read UGA as a codon for tryptophan rather than as a stop signal. Furthermore, AGA and AGG are read as stop signals rather than as codons for arginine, and AUA is read as a codon for methionine instead of isoleucine. Mitochondria of other species, such as those of yeast, also have genetic codes that differ slightly from the standard one. The genetic code of mitochondria can differ from that of the rest of the cell because mitochondrial DNA encodes a distinct set of tRNAs. Do any cellular protein-synthesizing systems deviate from the standard genetic code? Ciliated protozoa differ from most organisms in reading UAA and UAG as codons for amino acids rather than as stop signals; UGA is their sole termination signal. Thus, *the genetic code is nearly but not absolutely universal*. Variations clearly exist in mitochondria and in species, such as ciliates, that branched off very early in eukaryotic evolution. It is interesting to note that two of the codon reassignments in human mitochondria diminish the information content of the third base of the triplet (e.g., both AUA and AUG specify methionine). Most variations from the standard genetic code are in the direction of a simpler code.

Why has the code remained nearly invariant through billions of years of evolution, from bacteria to human beings? A mutation that altered the reading of mRNA would change the amino acid sequence of most, if not all, proteins synthesized by that particular organism. Many of these changes would undoubtedly be deleterious, and so there would be strong selection against a mutation with such pervasive consequences.

MOST EUKARYOTIC GENES ARE MOSAICS OF INTRONS AND EXONS

In bacteria, polypeptide chains are encoded by a continuous array of triplet codons in DNA. For many years, genes in higher organisms also were assumed to be continuous. This view was unexpectedly shattered in 1977, when investigators in several laboratories discovered that several genes are *discontinuous*. The mosaic nature of eukaryotic genes was revealed by electron microscopic studies of hybrids formed between mRNA and a segment of DNA containing the corresponding gene. For example, the gene for the β chain of hemoglobin is interrupted within its amino acid-coding sequence by a long *intervening sequence* of 550 base pairs and a short one of 120 base pairs. Thus, the *β-globin gene is split into three coding sequences.*

RNA PROCESSING GENERATES MATURE RNA

At what stage in gene expression are intervening sequences removed? Newly synthesized RNA chains (pre-mRNA) isolated from nuclei are much larger than the mRNA molecules derived from them: in the case of β-globin RNA, the former sediment at 15S in zonal centrifugation experiments and the latter at 9S. In fact, the primary transcript of the β-globin gene contains two regions that are not present in the mRNA. *These intervening sequences in the 15S primary transcript are excised, and the coding sequences are simultaneously linked by a precise splicing enzyme to form the mature 9S mRNA.* Regions that are removed from the primary transcript are called *introns* (for *int*ervening sequences), whereas those that are retained in the mature RNA are called *exons* (for *ex*pressed regions). A common feature in the expression of split genes is that their exons are ordered in the same sequence in mRNA as in DNA. Thus, split genes, like continuous genes, are colinear with their polypeptide products.

Splicing is a facile complex operation that is carried out by *spliceosomes,* which are assemblies of proteins and small RNA molecules. This enzymatic machinery recognizes signals in the nascent RNA that specify the splice sites. *Introns nearly always begin with GU and end with an AG that is preceded by a pyrimidine-rich tract. This consensus sequence is part of the signal for splicing.*

MANY EXONS ENCODE PROTEIN DOMAINS

Most genes of higher eukaryotes, such as birds and mammals, are split. Lower eukaryotes, such as yeast, have a much higher proportion of continuous genes. In prokaryotes, split genes are extremely rare. Have introns been inserted into genes in the evolution of higher organisms? Or have introns been removed from genes to form the streamlined genomes of prokaryotes and simple eukaryotes? Comparisons of the DNA sequences of genes encoding proteins that are highly conserved in evolution suggest that *introns were present in ancestral genes and were lost in the evolution of organisms that have become optimized for very rapid growth, such as prokaryotes.* The positions of introns in some genes are at least 1 billion years old. Furthermore, a common mechanism of splicing developed before the divergence of fungi, plants, and vertebrates, as shown by the finding that mammalian cell extracts can splice yeast RNA. *Many exons encode discrete structural and functional units of proteins.* An attractive hypothesis is that *new proteins arose in evolution by the rearrangement of exons encoding discrete structural elements, binding sites, and catalytic sites,* a process called *exon shuffling.* Because it preserves functional units but allows them to interact in new ways, exon shuffling is a rapid and efficient means of generating novel genes. Introns are extensive regions in which DNA can break and recombine with no deleterious effect on encoded proteins. In contrast, the exchange of sequences between different exons usually leads to loss of function.

Another advantage conferred by split genes is the potentiality for generating a series of related proteins by splicing a nascent RNA transcript in different ways. For example, a precursor of an antibody-producing cell forms an antibody that is anchored in the cell's plasma membrane. Stimulation of such a cell by a specific foreign antigen that is recognized by the attached antibody leads to cell differentiation and proliferation. The activated antibody-producing cells then splice their nascent RNA transcript in an alternative manner to form soluble antibody molecules that are secreted rather than retained on the cell surface. We see here a clear-cut example of a benefit conferred by the complex arrangement of introns and exons in higher organisms. Alternative splicing is a facile means of forming a set of proteins that are variations of a basic motif according to a developmental programme without requiring a gene for each protein.

Chapter 7

Exploring Genes

Recombinant DNA technology has revolutionized biochemistry since it came into being in the 1970s. The genetic endowment of organisms can now be precisely changed in designed ways. Recombinant DNA technology is a fruit of several decades of basic research on DNA, RNA, and viruses. It depends, first, on having enzymes that can cut, join, and replicate DNA and reverse transcribe RNA. Restriction enzymes cut very long DNA molecules into specific fragments that can be manipulated; DNA ligases join the fragments together. The availability of many kinds of restriction enzymes and DNA ligases makes it feasible to treat DNA sequences as modules that can be moved at will from one DNA molecule to another. Thus, recombinant DNA technology is based on nucleic acid enzymology.

A second foundation is the base-pairing language that allows complementary sequences to recognize and bind to each other. Hybridization with complementary DNA or RNA probes is a sensitive and powerful means of detecting specific nucleotide sequences. In recombinant DNA technology, base-pairing is used to construct new combinations of DNA as well as to detect and amplify particular sequences. This revolutionary technology is also critically dependent on our understanding of viruses, the ultimate parasites. Viruses efficiently deliver their own DNA (or RNA) into hosts, subverting them either to replicate the viral genome and produce viral proteins or to incorporate viral DNA into the host genome. Likewise, plasmids, which are accessory chromosomes found in bacteria, have been indispensable in recombinant DNA technology.

These new methods have wide-ranging benefits. Entire genomes, including the human genome, are being deciphered. New insights are emerging, for example, into the regulation of gene expression in cancer and development and the evolutionary history of proteins as well as organisms. New proteins can be created by altering genes in specific ways to provide detailed views into protein function. Clinically useful proteins, such as hormones, are now synthesized by recombinant DNA techniques. Crops are being generated to resist pests and harsh conditions. The new opportunities opened by recombinant DNA technology promise to have broad effects.

THE BASIC TOOLS OF GENE EXPLORATION

The rapid progress in biotechnology—indeed its very existence—is a result of a relatively few techniques.

1. *Restriction-enzyme analysis.* Restriction enzymes are precise, molecular scalpels that allow the investigator to manipulate DNA segments.
2. *Blotting techniques.* The Southern and Northern blots are used to separate and characterize DNA and RNA, respectively. The Western blot, which uses antibodies to characterize proteins.
3. *DNA sequencing.* The precise nucleotide sequence of a molecule of DNA can be determined. Sequencing has yielded a wealth of information concerning gene architecture, the control of gene expression, and protein structure.
4. *Solid-phase synthesis of nucleic acids.* Precise sequences of nucleic acids can be synthesized de novo and used to identify or amplify other nucleic acids.
5. *The polymerase chain reaction (PCR).* The polymerase chain reaction leads to a billionfold amplification of

a segment of DNA. One molecule of DNA can be amplified to quantities that permit characterization and manipulation. This powerful technique is being used to detect pathogens and genetic diseases, to determine the source of a hair left at the scene of a crime, and to resurrect genes from fossils.

A final tool, the use of which will be highlighted in the next chapter, is the computer. Without the computer, it would be impossible to catalog, access, and characterize the abundant information, especially DNA sequence information, that the techniques just outlined are rapidly generating.

RESTRICTION ENZYMES SPLIT DNA INTO SPECIFIC FRAGMENTS

Restriction enzymes, also called *restriction endonucleases,* recognize specific base sequences in double-helical DNA and cleave, at specific places, both strands of a duplex containing the recognized sequences. To biochemists, these exquisitely precise scalpels are marvelous gifts of nature. They are indispensable for analyzing chromosome structure, sequencing very long DNA molecules, isolating genes, and creating new DNA molecules that can be cloned. Werner Arber and Hamilton Smith discovered restriction enzymes, and Daniel Nathans pioneered their use in the late 1960s.

Restriction enzymes are found in a wide variety of prokaryotes. Their biological role is to cleave foreign DNA molecules. The cell's own DNA is not degraded, because the sites recognized by its own restriction enzymes are methylated. Many restriction enzymes recognize specific sequences of four to eight base pairs and hydrolyze a phosphodiester bond in each strand in this region. A striking characteristic of these cleavage sites is that they almost always possess *twofold rotational symmetry*. In other words, the recognized sequence is *palindromic,* or an inverted repeat, and the cleavage sites are symmetrically positioned. For example, the sequence recognized by a restriction enzyme from *Streptomyces achromogenes* is:

In each strand, the enzyme cleaves the C-G phosphodiester bond on the 3′ side of the symmetry axis. This symmetry reflects that of structures of the restriction enzymes themselves.

More than 100 restriction enzymes have been purified and characterized. Their names consist of a three-letter abbreviation for the host organism (e.g., *Eco* for *Escherichia coli, Hin* for *Haemophilus influenzae, Hae* for *Haemophilus aegyptius*) followed by a strain designation (if needed) and a roman numeral (if more than one restriction enzyme from the same strain has been identified). The specificities of several of these enzymes. Note that the cuts may be staggered or even.

Restriction enzymes are used to cleave DNA molecules into specific fragments that are more readily analysed and manipulated than the entire parent molecule. For example, the 5.1-kb circular duplex DNA of the tumor-producing SV40 virus is cleaved at 1 site by *Eco*RI, 4 sites by *Hpa*I, and 11 sites by *Hin*dIII. A piece of DNA produced by the action of one restriction enzyme can be specifically cleaved into smaller fragments by another restriction enzyme. The pattern of such fragments can serve as a *fingerprint* of a DNA molecule, as will be discussed shortly. Indeed, complex chromosomes containing hundreds of millions of base pairs can be mapped by using a series of restriction enzymes.

RESTRICTION FRAGMENTS CAN BE SEPARATED BY GEL ELECTROPHORESIS AND VISUALIZED

Small differences between related DNA molecules can be readily detected because their restriction fragments can be separated and displayed by gel electrophoresis. In many types of gels, the electrophoretic mobility of a DNA fragment is inversely proportional to the logarithm of the number of base pairs, up to a certain limit. Polyacrylamide gels are used to separate fragments containing about as many as 1000 base pairs, whereas more porous agarose gels are used to resolve mixtures of larger fragments (about as many as 20 kb). An important feature of these gels is their high resolving power.

In certain kinds of gels, *fragments differing in length by just one nucleotide of several hundred can be distinguished.* Moreover, entire chromosomes containing millions of nucleotides can be separated on agarose gels by applying pulsed electric fields (pulsed-field gel electrophoresis, PFGE) in different directions. This technique depends on the differential stretching and relaxing of large DNA molecules as an electric field is turned off and on at short intervals. Bands or spots of radioactive DNA in gels can be visualized by autoradiography. Alternatively, a gel can be stained with ethidium bromide, which fluoresces an intense orange when bound to double-helical DNA molecule. A band containing only 50 ng of DNA can be readily seen.

A restriction fragment containing a specific base sequence can be identified by hybridizing it with a labeled complementary DNA strand. A mixture of restriction fragments is separated by electrophoresis through an agarose gel, denatured to form single-stranded DNA, and transferred to a nitrocellulose sheet. The positions of the DNA fragments in the gel are preserved on the nitrocellulose sheet, where they are exposed to a ^{32}P-labeled single-stranded *DNA probe.* The probe hybridizes with a restriction fragment having a complementary sequence, and autoradiography then reveals the position of the restriction-fragment-probe duplex. A particular fragment in the midst of a million others can be readily identified in this way, like finding a needle in a haystack. This powerful technique is known as *Southern blotting* because it was devised by Edwin Southern. Similarly, RNA molecules can be separated by gel electrophoresis, and specific sequences can be identified by hybridization subsequent to their transfer to nitrocellulose. This analogous technique for the analysis of RNA has been whimsically termed *Northern blotting.* A further play on words accounts for the term *Western blotting,* which refers to a technique for detecting a particular protein by staining with specific antibody. Southern, Northern, and Western blots are also known respectively as *DNA, RNA,* and *protein blots.*

DNA IS USUALLY SEQUENCED BY CONTROLLED TERMINATION OF REPLICATION (SANGER DIDEOXY METHOD)

The analysis of DNA structure and its role in gene expression also have been markedly facilitated by the development of powerful techniques for the *sequencing* of DNA molecules. The key to DNA sequencing is the generation of DNA fragments whose length depends on the last base in the sequence. Collections of such fragments can be generated through the *controlled interruption of enzymatic replication,* a method developed by Frederick Sanger and coworkers. This technique has superseded alternative methods because of its simplicity. The same procedure is performed on four reaction mixtures at the same time. In all these mixtures, a DNA polymerase is used to make the complement of a particular sequence within a single-stranded DNA molecule. The synthesis is primed by a fragment, usually obtained by chemical synthetic methods described in figure that is complementary to a part of the sequence known from other studies. In addition to the four deoxyribonucleoside triphosphates (radioactively labeled), each reaction mixture contains a small amount of the 2′,3′ *-dideoxy analog* of *one* of the nucleotides, a different nucleotide for each reaction mixture. The incorporation of this analog blocks further growth of the new chain because it lacks the 3′ -hydroxyl terminus needed to form the next phosphodiester bond. The concentration of the dideoxy analog is low enough that chain termination will take place only occasionally. The polymerase will sometimes insert the correct nucleotide and other times the dideoxy analog, stopping the reaction. For instance, if the dideoxy analog of dATP is present, fragments of various lengths are produced, but all will be terminated by the dideoxy analog. Importantly, this dideoxy analog of dATP will be inserted only where a T was located in the DNA being sequenced. Thus, the fragments of different length will correspond to the positions of T. Four such sets of *chain-terminated fragments* (one for each dideoxy analog) then undergo electrophoresis, and the base sequence of the new DNA is read from the autoradiogram of the four lanes.

Fluorescence detection is a highly effective alternative to autoradiography. A fluorescent tag is attached to an oligonucleotide priming fragment—a differently coloured one in each of the four chain-terminating reaction mixtures (e.g., a blue emitter for termination at A and a red one for termination at C). The reaction mixtures are combined and subjected to electrophoresis together. The separated bands of DNA are then detected by their fluorescence as they emerge from the gel; the sequence of their colours directly gives the base sequence. Sequences of as many as 500 bases can be determined in this way. Alternatively, the dideoxy analogs can be labeled, each with a specific fluorescent label. When this method is used, all four terminators can be placed in a single tube, and only one reaction is necessary. Fluorescence detection is attractive because it eliminates the use of radioactive reagents and can be readily automated.

Sanger and coworkers determined the complete sequence of the 5386 bases in the DNA of the φ × 174 DNA virus in 1977, just a quarter century after Sanger's pioneering elucidation of the amino acid sequence of a protein. This accomplishment is a landmark in molecular biology because it revealed the total information content of a DNA genome. This tour de force was followed several years later by the determination of the sequence of human mitochondrial DNA, a double-stranded circular DNA molecule containing 16,569 base pairs. It encodes 2 ribosomal RNAs, 22 transfer RNAs, and 13 proteins. In recent years, the complete genomes of free-living organisms have been sequenced. The first such sequence to be completed was that of the bacterium *Haemophilus influenzae*. Its genome comprises 1,830,137 base pairs and encodes approximately 1740 proteins.

Many other bacterial and archaeal genomes have since been sequenced. The first eukaryotic genome to be completely sequenced was that of baker's yeast, *Saccharomyces cerevisiae*, which comprises approximately 12 million base pairs, distributed on 16 chromosomes, and encodes more than 6000 proteins. This achievement was followed by the first complete sequencing of the genome of a multicellular organism, the nematode *Caenorhabditis elegans*, which contains nearly 100 million base pairs. The human genome is considerably larger at more than 3 billion base pairs, but it has been essentially completely sequenced. *The ability to determine complete genome sequences has revolutionized biochemistry and biology.*

DNA PROBES AND GENES CAN BE SYNTHESIZED BY AUTOMATED SOLID-PHASE METHODS

DNA strands, like polypeptides can be synthesized by the sequential addition of activated monomers to a growing chain that is linked to an insoluble support. The activated monomers are protonated *deoxyribonucleoside 3′ -phosphoramidites*. In step 1, the 32 phosphorus atom of this incoming unit becomes joined to the 5′ oxygen atom of the growing chain to form a *phosphite triester*. The 5′ -OH group of the activated monomer is unreactive because it is blocked by a dimethoxytrityl (DMT) protecting group, and the 3′ -phosphoryl group is rendered unreactive by attachment of the β-cyanoethyl (βCE) group. Likewise, amino groups on the purine and pyrimidine bases are blocked.

Coupling is carried out under anhydrous conditions because water reacts with phosphoramidites. In step 2, the phosphite triester (in which P is trivalent) is oxidized by iodine to form a *phosphotriester* (in which P is pentavalent). In step 3, the DMT protecting group on the 5′ -OH of the growing chain is removed by the addition of dichloro-acetic acid, which leaves other protecting groups intact. The DNA chain is now elongated by one unit and ready for another cycle of addition. Each cycle takes only about 10 minutes and elongates more than 98per cent of the chains.

This solid-phase approach is ideal for the synthesis of DNA, as it is for polypeptides, because the desired product stays on the insoluble support until the final release step. All the reactions take place in a single vessel, and excess soluble reagents can be added to drive reactions to completion. At the end of each step, soluble reagents and by-products are washed away from the glass beads that bear the growing chains. At the end of the synthesis, NH_3 is added to remove all protecting groups and release the oligonucleotide from the solid support. Because elongation is never 100per cent complete, the new DNA chains are of diverse lengths—the desired chain is the longest one. The sample can be purified by high-pressure liquid chromatography or by electrophoresis on polyacrylamide gels. DNA chains of as many as 100 nucleotides can be readily synthesized by this automated method.

The ability to rapidly synthesize DNA chains of any selected sequence opens many experimental avenues. For example, synthesized oligonucleotide labeled at one end with ^{32}P or a fluorescent tag can be used to search for a complementary sequence in a very long DNA molecule or even in a genome consisting of many chromosomes. The use of labeled oligonucleotides as DNA probes is powerful and general. For example, a DNA probe that can base-pair to a known complementary sequence in a chromosome can serve as the starting point of an exploration of adjacent uncharted DNA. Such a probe can be used as a *primer* to initiate the replication of neighbouring DNA by DNA polymerase. One of the most exciting applications of the solid-phase approach is the *synthesis of new tailor-made genes*. New proteins with novel properties can now be produced in abundance by expressing synthetic genes. *Protein engineering* has become a reality.

SELECTED DNA SEQUENCES CAN BE GREATLY AMPLIFIED BY THE POLYMERASE CHAIN REACTION

In 1984, Kary Mullis devised an ingenious method called the *polymerase chain reaction (PCR)* for amplifying specific DNA sequences. Consider a DNA duplex consisting of a target sequence surrounded by nontarget DNA. Millions of the target sequences can be readily obtained by PCR if the flanking sequences of the target are known. PCR is carried out by adding the following components to a solution containing the target sequence:

1. A pair of primers that hybridize with the flanking sequences of the target,
2. All four deoxyribonucleoside triphosphates (dNTPs), and
3. A heat-stable DNA polymerase. A PCR cycle consists of three steps.
   a. *Strand separation.* The two strands of the parent DNA molecule are separated by heating the solution to 95°C for 15 s.
   b. *Hybridization of primers.* The solution is then abruptly cooled to 54°C to allow each primer to hybridize to a DNA strand. One primer hybridizes to the 3′ -end of the target on one strand, and the other primer hybridizes to the 32 end on the complementary target strand. Parent DNA duplexes do not form, because the primers are present in large excess. Primers are typically from 20 to 30 nucleotides long.
   c. *DNA synthesis.* The solution is then heated to 72°C, the optimal temperature for *Taq* DNA polymerase. This heat-stable polymerase comes from *Thermusaquaticus*, a thermophilic bacterium that lives in hot springs. The polymerase elongates both primers in the direction of the target sequence because DNA synthesis is in the 5′ -to-3′ direction. DNA synthesis takes place on both strands but extends beyond the target sequence.

These three steps—strand separation, hybridization of primers, and DNA synthesis—constitute one cycle of the PCR amplification and can be carried out repetitively just by changing the temperature of the reaction mixture. The thermostability of the polymerase makes it feasible to carry out PCR in a closed container; no reagents are added after the first cycle. The duplexes are heated to begin the second cycle, which produces four duplexes, and then the third cycle is initiated. At the end of the third cycle, two short strands appear that constitute only the target sequence—the sequence including and bounded by the primers. Subsequent cycles will amplify the target sequence exponentially. The larger strands increase in number arithmetically and serve as a source for the synthesis of more short strands. Ideally, after n cycles, this sequence is amplified 2^n-fold. The amplification is a millionfold after 20 cycles and a billionfold after 30 cycles, which can be carried out in less than an hour.

Several features of this remarkable method for amplifying DNA are noteworthy. First, the sequence of the target need not be known. All that is required is knowledge of the flanking sequences. Second, the target can be much larger than the primers. Targets larger than 10 kb have been amplified by PCR. Third, primers do not have to be perfectly matched to flanking sequences to amplify targets. With the use of primers derived from a gene of known sequence, it is possible to search for variations on the theme. In this way, families of genes are being discovered by PCR. Fourth, PCR is highly specific because of the stringency of hybridization at high temperature (54°C). Stringency is the required closeness of the match between primer and target, which can be controlled by temperature and salt. At high temperatures, the only DNA that is amplified is that situated between primers that have hybridized. A gene constituting less than a millionth of the total DNA of a higher organism is accessible by PCR. Fifth, PCR is exquisitely sensitive. A single DNA molecule can be amplified and detected.

PCR IS A POWERFUL TECHNIQUE IN MEDICAL DIAGNOSTICS, FORENSICS, AND MOLECULAR EVOLUTION

PCR can provide valuable diagnostic information in medicine. Bacteria and viruses can be readily detected with the use of specific primers. For example, PCR can reveal the presence of human immunodeficiency virus in people who have not mounted an immune response to this pathogen and would therefore be missed with an antibody assay. Finding *Mycobacterium tuberculosis* bacilli in tissue specimens is slow and laborious. With PCR, as few as 10 tubercle bacilli per million human cells can be readily detected. PCR is a promising method for the early detection of certain cancers. This technique can identify mutations of certain growth-control genes, such as the *ras* genes. The capacity to greatly amplify selected regions of DNA can also be highly informative in monitoring cancer chemotherapy. Tests using PCR can detect when cancerous cells have been eliminated and treatment can be stopped; they can also detect a relapse and the need to immediately resume treatment. PCR is ideal for detecting leukemias caused by chromosomal rearrangements.

PCR is also having an effect in forensics and legal medicine. An individual DNA profile is highly distinctive because many genetic loci are highly variable within a population. For example, variations at a specific one of these locations determines a person's HLA type (human leukocyte antigen type); organ transplants are rejected when the HLA types of the donor and recipient are not sufficiently matched. PCR amplification of multiple genes is being used to establish biological parentage in disputed paternity and immigration cases. Analyses of blood stains and semen samples by PCR have implicated guilt or innocence in numerous assault and rape cases. The root of a single shed hair found at a crime scene contains enough DNA for typing by PCR.

DNA is a remarkably stable molecule, particularly when relatively shielded from air, light, and water. Under such circumstances, large fragments of DNA can remain intact for thousands of years or longer. PCR provides an ideal method for amplifying such ancient DNA molecules so that they can be detected and characterized. PCR can also be used to amplify DNA from microorganisms that have not yet been isolated and cultured. As will be discussed in the next chapter, sequences from these PCR products can be sources of considerable insight into evolutionary relationships between organisms.

RECOMBINANT DNA TECHNOLOGY HAS REVOLUTIONIZED ALL ASPECTS OF BIOLOGY

The pioneering work of Paul Berg, Herbert Boyer, and Stanley Cohen in the early 1970s led to the development of recombinant DNA technology, which has permitted biology to move from an exclusively analytical science to a synthetic one. New combinations of unrelated genes can be constructed in the laboratory by applying recombinant DNA techniques. These novel combinations can be cloned—amplified manyfold—by introducing them into suitable cells, where they are replicated by the DNA-synthesizing machinery of the host. The inserted genes are often transcribed and translated in their new setting. What is most striking is that the genetic endowment of the host can be permanently altered in a designed way.

RESTRICTION ENZYMES AND DNA LIGASE ARE KEY TOOLS IN FORMING RECOMBINANT DNA MOLECULES

Let us begin by seeing how novel DNA molecules can be constructed in the laboratory. A DNA fragment of interest is covalently joined to a DNA *vector*. The essential feature of a vector is that it can replicate autonomously in an appropriate host. *Plasmids* (naturally occurring circles of DNA that act as accessory chromosomes in bacteria) and bacteriophage λ, a virus, are choice vectors for cloning in *E. coli*. The vector can be prepared for accepting a new DNA fragment by cleaving it at a single specific site with a restriction enzyme. For example, the plasmid pSC101, a 9.9-kb double-helical circular DNA molecule, is split at a unique site by the *Eco*RI restriction enzyme. The staggered cuts made by this enzyme produce *complementary single-stranded ends*, which have specific affinity for each other and hence are known as *cohesive* or *sticky ends*. Any DNA fragment can be inserted into this plasmid if it has the same cohesive ends. Such a fragment can be prepared from a larger piece of DNA by using the same restriction enzyme as was used to open the plasmid DNA.

The single-stranded ends of the fragment are then complementary to those of the cut plasmid. The DNA fragment and the cut plasmid can be annealed and then joined by *DNA ligase,* which catalyzes the formation of a phosphodiester bond at a break in a DNA chain. DNA ligase requires a free 3′-hydroxyl group and a 5′-phosphoryl group. Furthermore, the chains joined by ligase must be in a double helix. An energy source such as ATP or NAD^+ is required for the joining reaction.

This cohesive-end method for joining DNA molecules can be made general by using a *short, chemically synthesized DNA linker* that can be cleaved by restriction enzymes. First, the linker is covalently joined to the ends of a DNA fragment or vector. For example, the 5′ ends of a decameric linker and a DNA molecule are phosphorylated by polynucleotide kinase and then joined by the ligase from T4 phage. This ligase can form a covalent bond between blunt-ended (flush-ended) double-helical DNA molecules. Cohesive ends are produced when these terminal extensions are cut by an appropriate restriction enzyme. Thus, *cohesive ends corresponding to a particular restriction enzyme can be added to virtually any DNA molecule.* We see here the fruits of combining enzymatic and synthetic chemical approaches in crafting new DNA molecules.

PLASMIDS AND LAMBDA PHAGE ARE CHOICE VECTORS FOR DNA CLONING IN BACTERIA

Many plasmids and bacteriophages have been ingeniously modified to enhance the delivery of recombinant DNA molecules into bacteria and to facilitate the selection of bacteria harboring these vectors. Plasmids are circular duplex DNA molecules occurring naturally in some bacteria and ranging in size from 2 to several hundred kilobases. They carry genes for the inactivation of antibiotics, the production of toxins, and the breakdown of natural products. These *accessory chromosomes* can replicate independently of the host chromosome. In contrast with the host genome, they are dispensable under certain conditions. A bacterial cell may have no plasmids at all or it may house as many as 20 copies of a plasmid.

PBR322 PLASMID

One of the most useful plasmids for cloning is *pBR322,* which contains genes for resistance to tetracycline and ampicillin (an antibiotic like penicillin). Different endonucleases can cleave this plasmid at a variety of unique sites, at which DNA fragments can be inserted. Insertion of DNA at the *Eco*RI restriction site does not alter either of the genes for antibiotic resistance. However, insertion at the *Hind*III, *Sal*I, or *Bam*HI restriction site inactivates the gene for tetracycline resistance, an effect called *insertional inactivation.* Cells containing pBR322 with a DNA insert at one of these restriction sites are resistant to ampicillin but sensitive to tetracycline, and so they can be readily *selected.* Cells that failed to take up the vector are sensitive to both antibiotics, whereas cells containing pBR322 without a DNA insert are resistant to both.

LAMBDA (Λ) PHAGE

Another widely used vector, *λ phage,* enjoys a choice of life styles: this bacteriophage can destroy its host or it can become part of its host. In the *lytic pathway,* viral functions are fully expressed: viral DNA and proteins are quickly produced and packaged into virus particles, leading to the lysis (destruction) of the host cell and the sudden appearance of about 100 progeny virus particles, or *viricns.* In the *lyso-genic pathway,* the phage DNA becomes inserted into the host-cell genome and can be replicated together with host-cell DNA for many generations, remaining inactive. Certain environmental changes can trigger the expression of this dormant viral DNA, which leads to the formation of progeny virus and lysis of the host. Large segments of the 48-kb DNA of l phage are not essential for productive infection and can be replaced by foreign DNA, thus making λ phage an ideal vector.

Mutant λ phages designed for cloning have been constructed. An especially useful one called λgt-λβ contains only two *Eco*RI cleavage sites instead of the five normally present. After cleavage, the middle segment of this λ DNA molecule can be removed. The two remaining pieces of DNA (called arms) have a combined length equal to 72per cent of a normal genome length. This amount of DNA is too little to be packaged into a λ particle, because only DNA measuring from 75per cent to 105per cent of a normal genome in length can be readily packaged. However, *a suitably long DNA insert (such as 10 kb) between the two ends of λ DNA enables such a recombinant DNA molecule (93per cent of normal length) to be packaged.* Nearly all infective l particles formed in this way will contain an inserted piece of foreign DNA. Another advantage of using these modified viruses as vectors is that they enter bacteria much more easily than do plasmids. Among the variety of λ mutants that have been constructed for use as cloning vectors, one of them, called a *cosmid,* is essentially a hybrid of λ phage and a plasmid that can serve as a vector for large DNA inserts (as large as 45 kb).

M13 PHAGE

Another very useful vector for cloning DNA, *M13 phage* is especially useful for sequencing the inserted DNA. This filamentous virus is 900 nm long and only 9 nm wide. Its 6.4-kb single-stranded circle of DNA is protected by a coat of 2710 identical protein subunits. M13 enters *E. coli* through the bacterial sex pilus, a protein appendage that permits the transfer of DNA between bacteria. The single-stranded DNA in the virus particle [called the (+) strand] is replicated through an intermediate circular double-stranded replicative form (RF) containing (+) and (-) strands. Only the (+) strand is packaged into new virus particles. About a thousand progeny M13 are produced per generation. A striking feature of M13 is that it does not kill its bacterial host. Consequently, large quantities of M13 can be grown and easily harvested (1 gram from 10 liters of culture fluid).

An M13 vector is prepared for cloning by cutting its circular RF DNA at a single site with a restriction enzyme. The cut is made in a *polylinker* region that contains a series of closely spaced recognition sites for restriction enzymes; only one of each such sites is present in the vector. A double-stranded foreign DNA fragment produced by cleavage with the same restriction enzyme is then ligated to the cut vector. The foreign DNA can be inserted in two different orientations because the ends of both DNA molecules are the same. Hence, half the new (+) strands packaged into virus particles will contain one of the strands of the foreign DNA, and half will contain the other strand. Infection of *E. coli* by a single virus particle will yield a large amount of single-stranded M13 DNA containing the same strand of the foreign DNA. DNA cloned into M13 can be easily sequenced. An oligonucleotide that hybridizes adjacent to the polylinker region is used as a primer for sequencing the insert. This oligomer is called a *universal sequencing primer* because it can be used to sequence *any* insert. M13 is ideal for sequencing but not for long-term propagation of recombinant DNA, because inserts longer than about 1 kb are not stably maintained.

SPECIFIC GENES CAN BE CLONED FROM DIGESTS OF GENOMIC DNA

Ingenious cloning and selection methods have made feasible the isolation of a specific DNA segment several kilobases long out of a genome containing more than 3×10^6 kb. Let us see how a gene that is present just once in a human genome can be cloned. A sample containing many molecules of total genomic DNA is first mechanically sheared or partly digested by restriction enzymes into large fragments. This nearly random population of overlapping DNA fragments is then separated by gel electrophoresis to isolate a set about 15 kb long. Synthetic linkers are attached to the ends of these fragments, cohesive ends are formed, and the fragments are then inserted into a vector, such as λ phage DNA, prepared with the same cohesive ends. *E. coli* bacteria are then infected with these recombinant phages. The resulting lysate contains fragments of human DNA housed in a sufficiently large number of virus particles to ensure that nearly the entire genome is represented. These phages constitute a *genomic library*. Phages can be propagated indefinitely, and so the library can be used repeatedly over long periods.

This genomic library is then screened to find the very small portion of phages harboring the gene of interest. For the human genome, a calculation shows that a 99per cent probability of success requires screening about 500,000 clones; hence, a very rapid and efficient screening process is essential. Rapid screening can be accomplished by DNA hybridization.

A dilute suspension of the recombinant phages is first plated on a lawn of bacteria. Where each phage particle has landed and infected a bacterium, a *plaque* containing identical phages develops on the plate. A replica of this master plate is then made by applying a sheet of nitrocellulose. Infected bacteria and phage DNA released from lysed cells adhere to the sheet in a pattern of spots corresponding to the plaques. Intact bacteria on this sheet are lysed with NaOH, which also serves to denature the DNA so that it becomes accessible for hybridization with a ^{32}P-labeled probe. *The presence of a specific DNA sequence in a single spot on the replica can be detected by using a radioactive complementary DNA or RNA molecule as a probe.* Autoradiography then reveals the positions of spots harboring recombinant DNA. The corresponding plaques are picked out of the intact master plate and grown. A single investigator can readily screen a million clones in a day.

This method makes it possible to isolate virtually any gene, *provided that a probe is available.* How does one obtain a specific probe? One approach is to *start with the corresponding mRNA from cells in which it is abundant.* For example, precursors of red blood cells contain large amounts of mRNA for hemoglobin, and plasma cells are rich in mRNAs for antibody molecules. The mRNAs from these cells can be fractionated by size to enrich for the one of interest. As will be described shortly, a DNA complementary to this mRNA can be synthesized in vitro and cloned to produce a highly specific probe.

Alternatively, *a probe for a gene can be prepared if part of the amino acid sequence of the protein encoded by the gene is known.* A problem arises because a given peptide sequence can be encoded by a number of oligonucleotides. Thus, for this purpose, peptide sequences containing tryptophan and methionine are preferred, because these amino acids are specified by a single codon, whereas other amino acid residues have between two and six codons.

All the DNA sequences (or their complements) that encode the selected peptide sequence are synthesized by the solid-phase method and made radioactive by phosphorylating their 52 ends with ^{32}P from [^{32}P]-ATP. The replica plate is exposed to a mixture containing all these probes and autoradiographed to identify clones with a complementary DNA sequence. Positive clones are then sequenced to determine which ones have a sequence matching that of the protein of interest. Some of them may contain the desired gene or a significant segment of it.

LONG STRETCHES OF DNA CAN BE EFFICIENTLY ANALYSED BY CHROMOSOME WALKING

A typical genomic DNA library housed in λ phage vectors consists of DNA fragments about 15 kb long. However, many eukaryotic genes are much longer—for example, the *dystrophin* gene, which is mutated in Duchenne muscular dystrophy, is 2000 kb long. How can such long stretches of DNA be analysed? The development of cosmids helped because these chimeras of plasmids and λ phages can house 45-kb inserts. Much larger pieces of DNA can now be propagated in *bacterial artificial chromosomes (BACs)* or *yeast artificial chromosomes (YACs)*. YACs contain a centromere, an *autonomous replicating sequence* (ARS, where replication begins), a pair of telomeres (normal ends of eukaryotic chromosomes), selectable marker genes, and a cloning site. Genomic DNA is partly digested by a restriction endonuclease that cuts, on the average, at distant sites. The fragments are then separated by pulsed-field gel electrophoresis, and the large ones (~ 450 kb) are eluted and ligated into YACs. Artificial chromosomes bearing inserts ranging from 100 to 1000 kb are efficiently replicated in yeast cells.

Equally important in analyzing large genes is the capacity to scan long regions of DNA. The principle technique for this purpose makes use of overlaps in the library fragments. The fragments in a cosmid or YAC library are produced by random cleavage of many DNA molecules, and so some of the fragments overlap one another. Suppose that a fragment containing region *A* selected by hybridization with a complementary probe *A′* also contains region *B*. A new probe *B′* can be prepared by cleaving this fragment between regions *A* and *B* and subcloning region *B*. If the library is screened again with probe *B′*, new fragments containing region *B* will be found. Some will contain a previously unknown region C. Hence, we now have information about a segment of DNA encompassing regions *A*, *B*, and C. This process of subcloning and rescreening is called *chromosome walking*. Long stretches of DNA can be analysed in this way, provided that each of the new probes is complementary to a unique region.

MANIPULATING THE GENES OF EUKARYOTES

Eukaryotic genes, in a simplified form, can be introduced into bacteria, and the bacteria can be used as factories to produce a desired protein product. It is also possible to introduce DNA into higher organisms. In regard to animals, this ability provides a valuable tool for examining gene action, and it will be the basis of gene therapy. In regard to plants, introduced genes may make a plant resistant to pests or capable of growing in harsh conditions or able to carry greater quantities of essential nutrients. The manipulation of eukaryotic genes holds much promise for medical and agricultural benefits, but it is also the source of controversy.

COMPLEMENTARY DNA PREPARED FROM MRNA CAN BE EXPRESSED IN HOST CELLS

How can mammalian DNA be cloned and expressed by *E. coli?* Recall that most mammalian genes are mosaics of introns and exons. These interrupted genes cannot be expressed by bacteria, which lack the machinery to splice introns out of the primary transcript. However, this difficulty can be circumvented by causing bacteria to take up recombinant DNA that is complementary to mRNA. For example, proinsulin, a precursor of insulin, is synthesized by bacteria harboring plasmids that contain DNA complementary to mRNA for proinsulin. Indeed, bacteria produce much of the insulin used today by millions of diabetics.

The key to forming *complementary DNA (cDNA)* is the enzyme *reverse transcriptase.* A retrovirus uses this enzyme to form a DNA-RNA hybrid in replicating its genomic RNA. Reverse transcriptase synthesizes a DNA strand complementary to an RNA template if it is provided with a DNA primer that is base-paired to the RNA and contains a free 3′ -OH group. We can use a simple sequence of linked thymidine [oligo(T)] residues as the primer. This oligo(T) sequence pairs with the poly(A) sequence at the 3′ end of most eukaryotic mRNA molecules. The reverse transcriptase then synthesizes the rest of the cDNA strand in the presence of the four deoxyribonucleoside triphosphates. The RNA strand of this RNA-DNA hybrid is subsequently hydrolyzed by raising the pH. Unlike RNA, DNA is resistant to alkaline hydrolysis. The single-stranded DNA is converted into double-stranded DNA by creating another primer site. The enzyme *terminal transferase* adds nucleotides—for instance, several residues of dG—to the 3′ end of DNA. Oligo(dC) can bind to dG residues and prime the synthesis of the second DNA strand. Synthetic linkers can be added to this double-helical DNA for ligation to a suitable vector. Complementary DNA for all mRNA that a cell contains can be made, inserted into vectors, and then inserted into bacteria. Such a collection is called a *cDNA library*.

Complementary DNA molecules can be inserted into vectors that favour their efficient expression in hosts such as *E. coli.* Such plasmids or phages are called *expression vectors*. To maximize transcription, the cDNA is inserted into the vector in the correct reading frame near a strong bacterial promoter site. In addition, these vectors ensure efficient translation by encoding a ribosome-binding site on the mRNA near the initiation codon. Clones of cDNA can be screened on the basis of their capacity to direct the synthesis of a foreign protein in bacteria. A radioactive antibody specific for the protein of interest can be used to identify colonies of bacteria that harbor the corresponding cDNA vector. Spots of bacteria on a replica plate are lysed to release proteins, which bind to an applied nitrocellulose filter. A ^{125}I-labeled antibody specific for the protein of interest is added, and autoradiography reveals the location of the desired colonies on the master plate. This immunochemical screening approach can be used whenever a protein is expressed and corresponding antibody is available.

GENE-EXPRESSION LEVELS CAN BE COMPREHENSIVELY EXAMINED

Most genes are present in the same quantity in every cell—namely, one copy per haploid cell or two copies per diploid cell. However, the level at which a gene is expressed, as indicated by mRNA quantities, can vary widely, ranging from no expression to hundreds of mRNA copies per cell. Gene-expression patterns vary from cell type to cell type, distinguishing, for example, a muscle cell from a nerve cell. Even within the same cell, gene-expression levels may vary as the cell responds to changes in physiological circumstances.

Using our knowledge of complete genome sequences, it is now possible to analyse the pattern and level of expression of all genes in a particular cell or tissue. One of the most powerful methods developed to date for this purpose is based on hybridization. High-density arrays of oligonucleotides, called *DNA microarrays* or *gene chips,* can be constructed either through lightdirected chemical synthesis carried out with photolithographic microfabrication techniques used in the semiconductor industry or by placing very small dots of oligonucleotides or cDNAs on a solid support such as a microscope slide. Fluorescently labeled cDNA is hybridized to the chip to reveal the expression level for each gene, identifiable by its known location on the chip.

The intensity of the fluorescent spot on the chip reveals the extent of transcription of a particular gene. DNA chips have been prepared that contain oligonucleotides complementary to all known open reading frames, 6200 in number, within the yeast genome. An analysis of mRNA pools with the use of these chips revealed, for example, that approximately 50per cent of all yeast genes are expressed at steady-state levels of between 0.1 and 1.0 mRNA copy per cell. This method readily detected variations in expression levels displayed by specific genes under different growth conditions. These tools will continue to grow in power as genome sequencing efforts continue.

NEW GENES INSERTED INTO EUKARYOTIC CELLS CAN BE EFFICIENTLY EXPRESSED

Bacteria are ideal hosts for the amplification of DNA molecules. They can also serve as factories for the production of a wide range of prokaryotic and eukaryotic proteins. However, bacteria lack the necessary enzymes to carry out posttranslational modifications such as the specific cleavage of polypeptides and the attachment of carbohydrate units. Thus, many eukaryotic genes can be correctly expressed only in eukaryotic host cells. The introduction of recombinant DNA molecules into cells of higher organisms can also be a source of insight into how their genes are organized and expressed. How are genes turned on and off in embryological development? How does a fertilized egg give rise to an organism with highly differentiated cells that are organized in space and time? These central questions of biology can now be fruitfully approached by expressing foreign genes in mammalian cells.

Recombinant DNA molecules can be introduced into animal cells in several ways. In one method, foreign DNA molecules precipitated by calcium phosphate are taken up by animal cells. A small fraction of the imported DNA becomes stably integrated into the chromosomal DNA. The efficiency of incorporation is low, but the method is useful because it is easy to apply. In another method, DNA is *microinjected* into cells. A fine-tipped (0.1-μm-diameter) glass micropipet containing a solution of foreign DNA is inserted into a nucleus. A skilled investigator can inject hundreds of cells per hour. About 2per cent of injected mouse cells are viable and contain the new gene. In a third method, *viruses* are used to bring new genes into animal cells. The most effective vectors are *retroviruses*. Retroviruses replicate through DNA intermediates, the reverse of the normal flow of information. A striking feature of the life cycle of a retrovirus is that the double-helical DNA form of its genome, produced by the action of reverse transcriptase, becomes randomly incorporated into host chromosomal DNA. This DNA version of the viral genome, called *proviral DNA*, can be efficiently expressed by the host cell and replicated along with normal cellular DNA. Retroviruses do not usually kill their hosts. Foreign genes have been efficiently introduced into mammalian cells by infecting them with vectors derived from *Moloney murine leukemia virus*, which can accept inserts as long as 6 kb. Some genes introduced by this retroviral vector into the genome of a transformed host cell are efficiently expressed.

Two other viral vectors are extensively used. *Vaccinia virus*, a large DNA-containing virus, replicates in the cytoplasm of mammalian cells, where it shuts down host-cell protein synthesis. *Baculovirus* infects insect cells, which can be conveniently cultured. Insect larvae infected with this virus can serve as efficient protein factories. Vectors based on these large-genome viruses have been engineered to express DNA inserts efficiently.

TRANSGENIC ANIMALS HARBOR AND EXPRESS GENES THAT WERE INTRODUCED INTO THEIR GERM LINES

Genetically engineered giant mice illustrate the expression of foreign genes in mammalian cells. Giant mice were produced by introducing the gene for rat growth hormone into a fertilized mouse egg. *Growth hormone (somatotropin)*, a 21-kd protein, is normally synthesized by the pituitary gland. A deficiency of this hormone produces dwarfism, and an excess leads to gigantism. The gene for rat growth hormone was placed on a plasmid next to the mouse metallothionein *promoter*. This promoter site is normally located on a chromosome, where it controls the transcription of *metallothionein*, a cysteine-rich protein that has high affinity for heavy metals. Metallothionein binds to and sequesters heavy metals, many of which are toxic for metabolic processes. The synthesis of this protective protein by the liver is induced by heavy-metal ions such as cadmium. Hence, if mice contain the new gene, its expression can be initiated by the addition of cadmium to the drinking water.

Several hundred copies of the plasmid containing the promoter and growth-hormone gene were microinjected into the male pronucleus of a fertilized mouse egg, which was then inserted into the uterus of a foster mother mouse. A number of mice that developed from such microinjected eggs contained the gene for rat growth hormone, as shown by Southern blots of their DNA. These *transgenic mice*, containing multiple copies (30 per cell) of the rat growth-hormone gene, grew much more rapidly than did control mice. In the presence of cadmium, the level of growth hormone in these mice was 500 times as high as in normal mice, and their body weight at maturity was twice normal. The foreign DNA had been transcribed and its five introns correctly spliced out to form functional mRNA. *These experiments strikingly demonstrate that a foreign gene under the control of a new promoter site can be integrated and efficiently expressed in mammalian cells.*

GENE DISRUPTION PROVIDES CLUES TO GENE FUNCTION

A gene's function can also be probed by inactivating the gene and looking for resulting abnormalities. Powerful methods have been developed for accomplishing *gene disruption* (also called *gene knockout*) in organisms such as yeast and mice. These methods rely on the process of *homologous recombination*. Through this process, regions of strong sequence similarity exchange segments of DNA. Foreign DNA inserted into a cell thus can disrupt any gene that is at least in part homologous by exchanging segments. Specific genes can be targeted if their nucleotide sequences are known.

For example, the gene knockout approach has been applied to the genes encoding gene regulatory proteins (also called transcription factors) that control the differentiation of muscle cells. When both copies of the gene for the regulatory protein *myogenin* are disrupted, an animal dies at birth because it lacks functional skeletal muscle. Microscopic inspection reveals that the tissues from which muscle normally forms contain precursor cells that have failed to differentiate fully. Heterozygous mice containing one normal myogenin gene and one disrupted gene appear normal, indicating that the level of gene expression is not essential for its function. Analogous studies have probed the function of many other genes to generate animal models for known human genetic diseases.

TUMOR-INDUCING PLASMIDS CAN BE USED TO INTRODUCE NEW GENES INTO PLANT CELLS

The common soil bacterium *Agrobacterium tumefaciens* infects plants and introduces foreign genes into plants cells. A lump of tumor tissue called a *crown gall* grows at the site of infection. Crown galls synthesize opines, a group of amino acid derivatives that are metabolized by the infecting bacteria. In essence, the metabolism of the plant cell is diverted to satisfy the highly distinctive appetite of the intruder. *Tumor-inducing plasmids (Ti plasmids)* that are carried by *Agrobacterium* carry instructions for the switch to the tumor state and the synthesis of opines. A small part of the Ti plasmid becomes integrated into the genome of infected plant cells; this 20-kb segment is called *T-DNA*.

Ti plasmid derivatives can be used as vectors to deliver foreign genes into plant cells. First, a segment of foreign DNA is inserted into the T-DNA region of a small plasmid through the use of restriction enzymes and ligases. This synthetic plasmid is added to *Agrobacterium* colonies harboring naturally occurring Ti plasmids. By recombination, Ti plasmids containing the foreign gene are formed. These Ti vectors hold great promise as tools for exploring the genomes of plant cells and modifying plants to improve their agricultural value and crop yield. However, they are not suitable for transforming all types of plants. Ti-plasmid transfer is effective with dicots (broad-leaved plants such as grapes) and a few kinds of monocots but not with economically important cereal monocots.

Foreign DNA can be introduced into cereal monocots as well as dicots by applying intense electric fields, a technique called *electroporation*. First, the cellulose wall surrounding plant cells is removed by adding cellulase; this treatment produces *protoplasts*, plant cells with exposed plasma membranes. Electric pulses are then applied to a suspension of protoplasts and plasmid DNA. Because high electric fields make membranes transiently permeable to large molecules, plasmid DNA molecules enter the cells. The cell wall is then allowed to reform, and the plant cells are again viable. Maize cells and carrot cells have been stably transformed in this way with the use of plasmid DNA that includes genes for resistance to antibiotics. Moreover, the transformed cells efficiently express the plasmid DNA. Electroporation is also an effective means of delivering foreign DNA into animal cells.

The most effective means of transforming plant cells is through the use of "gene guns," or bombardment-mediated transformation. DNA is coated onto 1-μm-diameter tungsten pellets, and these microprojectiles are fired at the target cells with a velocity greater than 400 m s^{-1}. Despite its apparent crudeness, this technique is proving to be the most effective way of transforming plants, especially important crop species such as soybean, corn, wheat, and rice. The gene-gun technique affords an opportunity to develop genetically modified organisms (GMOs) with beneficial characteristics. Such characteristics could include the ability to grow in poor soils, resistance to natural climatic variation, resistance to pests, and nutritional fortification. These crops might be most useful in developing countries. The use of genetically modified organisms is highly controversial at this point because of fears of unexpected side effects.

The first GMO to come to market was a tomato characterized by delayed ripening, rendering it ideal for shipment. Pectin is a polysaccharide that gives tomatoes their firmness and is naturally destroyed by the enzyme *polygalacturonase.* As pectin is destroyed, the tomatoes soften, making shipment difficult. DNA was introduced that disrupts the polygalacturonase gene. Less of the enzyme was produced, and the tomatoes stayed fresh longer. However, the tomato's poor taste hindered its commercial success.

NOVEL PROTEINS CAN BE ENGINEERED BY SITE-SPECIFIC MUTAGENESIS

Much has been learned about genes and proteins by analyzing mutated genes selected from the repertoire offered by nature. In the classic genetic approach, mutations are generated randomly throughout the genome, and those exhibiting a particular phenotype are selected. Analysis of these mutants then reveals which genes are altered, and DNA sequencing identifies the precise nature of the changes. *Recombinant DNA technology now makes it feasible to create specific mutations in vitro.*

PROTEINS WITH NEW FUNCTIONS CAN BE CREATED THROUGH DIRECTED CHANGES IN DNA

We can construct new genes with designed properties by making three kinds of directed changes: *deletions, insertions,* and *substitutions.*

DELETIONS

A specific deletion can be produced by cleaving a plasmid at two sites with a restriction enzyme and ligating to form a smaller circle. This simple approach usually removes a large block of DNA. A smaller deletion can be made by cutting a plasmid at a single site. The ends of the linear DNA are then digested with an exonuclease that removes nucleotides from both strands. The shortened piece of DNA is then ligated to form a circle that is missing a short length of DNA about the restriction site.

SUBSTITUTIONS: OLIGONUCLEOTIDE-DIRECTED MUTAGENESIS

Mutant proteins with single amino acid substitutions can be readily produced by *oligonucleotide-directed mutagenesis.*

Suppose that we want to replace a particular serine residue with cysteine. This mutation can be made if (1) we have a plasmid containing the gene or cDNA for the protein and (2) we know the base sequence around the site to be altered. If the serine of interest is encoded by TCT, we need to change the C to a G to get cysteine, which is encoded by TGT. This type of mutation is called a point mutation because only one base is altered. The key to this mutation is to prepare an oligonucleotide primer that is complementary to this region of the gene except that it contains TGT instead of TCT. The two strands of the plasmid are separated, and the primer is then annealed to the complementary strand. The mismatch of 1 base pair of 15 is tolerable if the annealing is carried out at an appropriate temperature. After annealing to the complementary strand, the primer is elongated by DNA polymerase, and the double-stranded circle is closed by adding DNA ligase. Subsequent replication of this duplex yields two kinds of progeny plasmid, half with the original TCT sequence and half with the mutant TGT sequence. Expression of the plasmid containing the new TGT sequence will produce a protein with the desired substitution of serine for cysteine at a unique site. We will encounter many examples of the use of oligonucleotide-directed mutagenesis to precisely alter regulatory regions of genes and to produce proteins with tailor-made features.

INSERTIONS: CASSETTE MUTAGENESIS

In another valuable approach, *cassette mutagenesis*, plasmid DNA is cut with a pair of restriction enzymes to remove a short segment. A synthetic double-stranded oligonucleotide (the *cassette*) with cohesive ends that are complementary to the ends of the cut plasmid is then added and ligated. Each plasmid now contains the desired mutation. It is convenient to introducc into the plasmid unique restriction sites spaced about 40 nucleotides apart so that mutations can be readily made anywhere in the sequence.

DESIGNER GENES

Novel proteins can also be created by splicing together gene segments that encode domains that are not associated in nature. For example, a gene for an antibody can be joined to a gene for a toxin to produce a chimeric protein that kills cells that are recognized by the antibody. These *immunotoxins* are being evaluated as anticancer agents. Entirely new genes can be synthesized de novo by the solid-phase method. Furthermore, noninfectious coat proteins of viruses can be produced in large amounts by recombinant DNA methods. They can serve as *synthetic vaccines* that are safer than conventional vaccines prepared by inactivating pathogenic viruses. A subunit of the hepatitis B virus produced in yeast is proving to be an effective vaccine against this debilitating viral disease.

RECOMBINANT DNA TECHNOLOGY HAS OPENED NEW VISTAS

Recombinant DNA technology has revolutionized the analysis of the molecular basis of life. Complex chromosomes are being rapidly mapped and dissected into units that can be manipulated and deciphered. The amplification of genes by cloning has provided abundant quantities of DNA for sequencing. New insights are emerging, as exemplified by the discovery of introns in eukaryotic genes. Central questions of biology, such as the molecular basis of development, are now being fruitfully explored. DNA and RNA sequences provide a wealth of information about evolution. Biochemists now move back and forth between gene and protein and feel at home in both areas of inquiry.

Analyses of genes and cDNA can reveal the existence of previously unknown proteins, which can be isolated and purified. Conversely, purification of a protein can be the starting point for the isolation and cloning of its gene or cDNA. Very small amounts of protein or nucleic acid suffice because of the sensitivity of recently developed microchemical techniques and the amplification afforded by gene cloning and the polymerase chain reaction. The powerful techniques of protein chemistry, nucleic acid chemistry, immunology, and molecular genetics are highly synergistic.

New kinds of proteins can be created by altering genes in specific ways. Site-specific mutagenesis opens the door to understanding how proteins fold, recognize other molecules, catalyze reactions, and process information. Large amounts of protein can be obtained by expressing cloned genes or cDNAs in bacteria or eukaryotic cells. Hormones, such as insulin, and antiviral agents, such as interferon, are being produced by bacteria. Tissue plasminogen activator, which is administered to a patient after a heart attack, is made in large quantities in mammalian cells. A new pharmacology, using proteins produced by recombinant DNA technology as drugs, is beginning to significantly alter the practice of medicine. Recombinant DNA technology is also providing highly specific diagnostic reagents, such as DNA probes for the detection of genetic diseases, infections, and cancers. Human gene therapy has been successfully initiated. White blood cells deficient in adenosine deaminase, an essential enzyme, are taken from patients and returned after being transformed in vitro to correct the genetic error. Agriculture, too, is benefiting from genetic engineering. Transgenic crops with increased resistance to insects, herbicides, and drought have been produced.

Chapter 8

Exploring Evolution

Like members of a human family, members of molecular families often have features in common. Such family resemblance is most easily detected by comparing three-dimensional structure, the aspect of a molecule most closely linked to function. Consider, for example, ribonuclease from cows, which was introduced in our consideration of protein folding. Comparing structures reveals that the three-dimensional structure of this protein and that of a human ribonuclease are quite similar. Although this similarity is not unexpected, given the similarity in biological function, similarities revealed by comparisons are sometimes surprising. For example, angiogenin, a protein identified on the basis of its ability to stimulate the growth of new blood vessels, also turns out to be structurally similar to ribonuclease—so similar that it is clear that both angiogenin and ribonuclease are members of the same protein family. Angiogenin and ribonuclease must have had a common ancestor at some earlier stage of evolution.

Unfortunately, three-dimensional structures have been determined for only a relatively small number of proteins. In contrast, gene sequences and the corresponding amino acid sequences are available for a great number of proteins, largely owing to the tremendous power of DNA cloning and sequencing techniques. Evolutionary relationships also are manifest in amino acid sequences. For example, comparison of the amino acid sequences of bovine ribonuclease and angiogenin reveals that 35per cent of amino acids in corresponding positions are identical. Is this level sufficiently high to ensure an evolutionary relationship? If not, what level is required? In this chapter, we shall examine the methods that are used to compare amino acid sequences and to deduce such evolutionary relationships.

Sequence-comparison methods have become a powerful tool in modern biochemistry. Sequence databases can be probed for matches to a newly elucidated sequence in order to identify related molecules. This information can often be a source of considerable insight into the function and mechanism of the newly sequenced molecule. When three-dimensional structures are available, they may be compared to confirm relationships suggested by sequence comparisons and to reveal others that are not readily detected at the level of sequence alone.

By examining the footprints present in modern protein sequences, the biochemist can become a molecular archeologist able to learn about events in the evolutionary past. Sequences comparisons can often reveal both pathways of evolutionary descent and estimated dates of specific evolutionary landmarks. This information can be used to construct evolutionary trees that trace the evolution of a particular protein or nucleic acid in many cases from Archaea and Bacteria through Eukarya, including human beings. Molecular evolution can also be studied experimentally. In some cases, DNA from fossils can be amplified by PCR methods and sequenced, giving a direct view into the past. In addition, investigators can observe molecular evolution taking place in the laboratory, through experiments based on nucleic acid replication. The results of such studies are revealing more about how evolution proceeds.

HOMOLOGS ARE DESCENDED FROM A COMMON ANCESTOR

The exploration of biochemical evolution consists largely of an attempt to determine how proteins, other molecules, and biochemical pathways have been transformed through time. The most fundamental relationship between two entities is *homology;* two molecules are said to be *homologous* if they have been derived from a common ancestor. Homologous molecules, or *homologs,* can be divided into two classes *Paralogs* are homologs that are present within one species. Paralogs often differ in their detailed biochemical functions. *Orthologs* are homologs that are present within different species and have very similar or identical functions. Understanding the homology between molecules can reveal the evolutionary history of the molecules as well as information about their function; if a newly sequenced protein is homologous to an already characterized protein, we have a strong indication of the new protein's biochemical function.

How can we tell whether two human proteins are paralogs or whether a yeast protein is the ortholog of a human protein? *Homology is often manifested by significant similarity in nucleotide or amino acid sequence and almost always manifested in three-dimensional structure.*

STATISTICAL ANALYSIS OF SEQUENCE ALIGNMENTS CAN DETECT HOMOLOGY

A significant sequence similarity between two molecules implies that they are likely to have the same evolutionary origin and, therefore, the same three-dimensional structure, function, and mechanism. Although both nucleic acid and protein sequences can be compared to detect homology, a comparison of protein sequences is much more effective for several reasons, most notably that proteins are built from 20 different building blocks, whereas RNA and DNA are synthesized from only 4 building blocks.

To illustrate sequence-comparison methods, let us consider a class of proteins called the *globins.* Myoglobin is a protein that binds oxygen in muscle, whereas hemoglobin is the oxygen-carrying protein in blood. Both proteins cradle a heme group, an iron-containing organic molecule that binds the oxygen. Each human hemoglobin molecule is composed of four heme-containing polypeptide chains, two identical α chains and two identical β chains. Here, we consider only the α chain. We wish to examine the similarity between the amino acid sequence of the human α chain and that of human myoglobin. To detect such similarity, methods have been developed for *sequence alignment.*

How can we tell where to align the two sequences? The simplest approach is to compare all possible juxtapositions of one protein sequence with another, in each case recording the number of identical residues that are aligned with one another. This comparison can be accomplished by simply sliding one sequence past the other, one amino acid at a time, and counting the number of matched residues.

For hemoglobin α and myoglobin, the best alignment reveals 23 sequence identities, spread throughout the central parts of the sequences. However, a nearby alignment showing 22 identities is nearly as good. In this alignment, the identities are concentrated toward the amino-terminal end of the sequences. The sequences can be aligned to capture most of the identities in *both* alignments by introducing a *gap* into one of the sequences. Such gaps must often be inserted to compensate for the insertions or deletions of nucleotides that may have taken place in the gene for one molecule but not the other in the course of evolution.

The use of gaps substantially increases the complexity of sequence alignment because, in principle, the insertion of gaps of arbitrary sizes must be considered throughout each sequence. However, methods have been developed for the insertion of gaps in the automatic alignment of sequences. These methods use scoring systems to compare different alignments, and they include penalties for gaps to prevent the insertion of an unreasonable number of them. Here is an example of such a scoring system: each identity between aligned sequences results in +10 points, whereas each gap introduced, regardless of size, results in -25 points. For the alignment shown in figure there are 38 identities and 1 gap, producing a score of ($38 \times 10 - 1 \times 25 = 355$). Overall, there are

38 matched amino acids in an average length of 147 residues; so the sequences are 25.9per cent identical. The next step is to ask, Is this precentage of identity significant?

THE STATISTICAL SIGNIFICANCE OF ALIGNMENTS CAN BE ESTIMATED BY SHUFFLING

The similarities in sequence in figure appear striking, yet there remains the possibility that a grouping of sequence identities has occurred by chance alone. How can we estimate the probability that a specific series of identities is a chance occurrence? To make such an estimate, the amino acid sequence in one of the proteins is "shuffled"—that is, randomly rearranged—and the alignment procedure is repeated. This process is repeated to build up a distribution showing, for each possible score, the number of shuffled sequences that received that score.

When this procedure is applied to the sequences of myoglobin and hemoglobin α, the authentic alignment clearly stands out. Its score is far above the mean for the alignment scores based on shuffled sequences. The odds of such a deviation occurring owing due to chance alone are approximately 1 in 10^{20}. Thus, we can comfortably conclude that the two sequences are genuinely similar; the simplest explanation for this similarity is that these sequences are homologous—that is, that the two molecules have descended by divergence from a common ancestor.

DISTANT EVOLUTIONARY RELATIONSHIPS CAN BE DETECTED THROUGH THE USE OF SUBSTITUTION MATRICES

The scoring scheme in figure assigns points only to positions occupied by identical amino acids in the two sequences being compared. No credit is given for any pairing that is not an identity. However, not all substitutions are equivalent. Some are structurally *conservative substitutions*, replacing one amino acid with another that is similar in size and chemical properties. Such conservative amino acid substitutions may have relatively minor effects on protein structure and can thus be tolerated without compromising function. In other substitutions, an amino acid replaces one that is dissimilar. Furthermore, some amino acid substitutions result from the replacement of only a single nucleotide in the gene sequence; whereas others require two or three replacements. Conservative and single-nucleotide substitutions are likely to be more common than are substitutions with more radical effects. How can we account for the type of substitution when comparing sequences? We can approach this problem by first examining the substitutions that have actually taken place in evolutionarily related proteins.

From the examination of appropriately aligned sequences, *substitution matrices* can be deduced. In these matrices, a large positive score corresponds to a substitution that occurs relatively frequently, whereas a large negative score corresponds to a substitution that occurs only rarely. The highest scores in this substitution matrix indicate that amino acids such as cysteine (C) and tryptophan (W) tend to be conserved more than those such as serine (S) and alanine (A). Furthermore, structurally conservative substitutions such as lysine (K) for arginine (R) and isoleucine (I) for valine (V) have relatively high scores. When two sequences are compared, each substitution is assigned a score based on the matrix. In addition, a gap penalty is often assigned according to the size of the gap. For example, the introduction of a gap lowers the alignment score by 12 points and the extension of an existing gap costs 2 points per residue. Using this scoring system, the alignment shown in figure receives a score of 115. In many regions, most substitutions are conservative (defined as those substitutions with scores greater than 0) and relatively few are strongly disfavoured types.

This scoring system detects homology between less obviously related sequences with greater sensitivity than would a comparison of identities only. Consider, for example, the protein leghemoglobin, an oxygen-binding protein found in the roots of some plants. The amino acid sequence of leghemoglobin from the herb lupine can be aligned with that of human myoglobin and scored by using either the simple scoring scheme based on identities only or the Blosum-62 scoring matrix. Repeated shuffling and scoring provides a distribution of alignment scores. Scoring based on identities only indicates that the odds of the alignment between myoglobin and leghemoglobin occurring by chance alone are 1 in 20. Thus, although the level of similarity suggests a relationship, there is a 5per cent chance that the similarity is accidental on the basis of this analysis. In contrast, users of the substitution matrix are able to incorporate the effects of conservative substitutions. From such an analysis, the odds of the alignment occurring by chance are calculated to be approximately 1 in 300. Thus, an analysis performed by using the substitution matrix reaches a much firmer conclusion about the evolutionary relationship between these proteins.

Experience with sequence analysis has led to the development of simpler rules of thumb. For sequences longer than 100 amino acids, sequence identities greater than 25per cent are almost certainly not the result of chance alone; such sequences are probably homologous. In contrast, if two sequences are less than 15per cent identical, pairwise comparison alone is unlikely to indicate statistically significant similarity. For sequences that are between 15per cent and 25per cent identical, further analysis is necessary to determine the statistical significance of the alignment. It must be emphasized that *the lack of a statistically significant degree of sequence similarity does not rule out homology*. The sequences of many proteins that have descended from common ancestors have diverged to such an extent that the relationship between the proteins can no longer be detected from their sequences alone. As we will see, such homologous proteins can often be detected by examining three-dimensional structures.

DATABASES CAN BE SEARCHED TO IDENTIFY HOMOLOGOUS SEQUENCES

When the sequence of a protein is first determined, comparing it with all previously characterized sequences can be a source of tremendous insight into its evolutionary relatives and, hence, its structure and function. *Indeed, an extensive sequence comparison is almost always the first analysis performed on a newly elucidated sequence.* The sequence alignment methods heretofore described are used to compare an individual sequence with all members of a database of known sequences.

In 1995, investigators reported the first complete sequence of the genome of a free-living organism, the bacterium *Haemophilus influenzae*. Of 1743 identified open reading frames 1007 (58per cent) could be linked by sequence-comparison methods to some protein of known function that had been previously characterized in another organism. An additional 347 open reading frames could be linked to sequences in the database for which no function had yet been assigned ("hypothetical proteins"). The remaining 389 sequences did not match any sequence present in the database at the time at which the *Haemophilus influenzae* sequence was completed. Thus, investigators were able to identify likely functions for more than half the proteins within this organism solely through the use of sequence-comparison methods.

EXAMINATION OF THREE-DIMENSIONAL STRUCTURE ENHANCES OUR UNDERSTANDING OF EVOLUTIONARY RELATIONSHIPS

Sequence comparison is a powerful tool for extending our knowledge of protein function and kinship. However, biomolecules generally function as intricate three-dimensional structures rather than as linear polymers. Mutations occur at the level of sequence, but the effects of the mutations are at the level of function, and function is directly related to tertiary structure. Consequently, to gain a deeper understanding of evolutionary relationships between proteins, we must examine three-dimensional structures, especially in conjunction with sequence information.

TERTIARY STRUCTURE IS MORE CONSERVED THAN PRIMARY STRUCTURE

Because three-dimensional structure is much more closely associated with function than is sequence, tertiary structure is more evolutionarily conserved than is primary structure. This conservation is apparent in the tertiary structures of the globins which are extremely similar even though the similarity between human myoglobin and lupine leghemoglobin is just barely detectable at the sequence level and that between human hemoglobin (α chain) and lupine leghemoglobin is not statistically significant (15.6per cent identity). This structural similarity firmly establishes that the framework that binds the heme group and facilitates the reversible binding of oxygen has been conserved over a long evolutionary period.

Anyone aware of the similar biochemical functions of hemoglobin, myoglobin, and leghemoglobin could expect the structural similarities. In a growing number of other cases, however, a comparison of three-dimensional structures has revealed striking similarities between proteins that were *not* expected to be related. A case in point is the protein actin, a major component of the cytoskeleton, and heat shock protein 70 (Hsp-70), which assists protein folding inside cells. These two proteins were found to be noticeably similar in structure despite only 15.6per cent sequence identity. On the basis of their three-dimensional structures, actin and Hsp-70 are paralogs. The level of structural similarity strongly suggests that, despite their different biological roles in modern organisms, these proteins descended from a common ancestor. As the three-dimensional structures of more proteins are determined, such unexpected kinships are being discovered with increasing frequency. The search for such kinships relies ever more frequently on computer-based search procedures that allow the three-dimensional structure of any protein to be compared with all other known structures.

KNOWLEDGE OF THREE-DIMENSIONAL STRUCTURES CAN AID IN THE EVALUATION OF SEQUENCE ALIGNMENTS

The sequence-comparison methods described thus far treat all positions within a sequence equally. However, examination of families of homologous proteins for which at least one three-dimensional structure is known has revealed that regions and residues critical to protein function are more strongly conserved than are other residues. For example, each type of globin contains a bound heme group with an iron atom at its centre. A histidine residue that interacts directly with this iron (residue 64 in human myoglobin) is conserved in all globins. After we have identified key residues or highly conserved sequences within a family of proteins, we can sometimes identify other family members even when the overall level of sequence similarity is below statistical significance. Thus, the generation of *sequence templates*—conserved residues that are structurally and functionally important and are characteristic of particular families of proteins—can be useful for recognizing new family members that might be undetectable by other means. A variety of other methods for sequence classification that take advantage of known three-dimensional structures also are being developed. Still other methods are able to identify relatively conserved residues within a family of homologous proteins, even without a known three-dimensional structure. These methods are proving to be powerful in identifying distant evolutionary relationships.

REPEATED MOTIFS CAN BE DETECTED BY ALIGNING SEQUENCES WITH THEMSELVES

More than 10per cent of all proteins contain sets of two or more domains that are similar to one another. The aforedescribed sequence search methods can often detect internally repeated sequences that have been characterized in other proteins. Where repeated units do not correspond to previously identified domains, their presence can be detected by attempting to align a given sequence with itself. This alignment is most easily visualized with the use of a *self-diagonal plot*. Here, the protein sequence is displayed on both the vertical and the horizontal axes, running from amino to carboxyl terminus; a dot is placed at each point in the space defined by the axes at which the amino acid directly below along the horizontal axis is the same as that directly across along the vertical axis. The central diagonal represents the sequence aligned with itself. Internal repeats are manifested as lines of dots parallel to the central diagonal, illustrated by the plot in figure prepared for the TATA-box-binding protein, a key protein in the initiation of gene transcription.

The statistical significance of such repeats can be tested by aligning the regions in question as if these regions were sequences from separate proteins. For the TATA-box-binding protein, the alignment is highly significant: 30per cent of the amino acids are identical over 90 residues. The estimated probability of such an alignment occurring by chance is 1 in 10^{13}. The determination of the three-dimensional structure of the TATA-box-binding protein confirmed the presence of repeated structures; the protein is formed of two nearly identical domains. The evidence is convincing that the gene encoding this protein evolved by duplication of a gene encoding a single domain.

CONVERGENT EVOLUTION: COMMON SOLUTIONS TO BIOCHEMICAL CHALLENGES

Thus far, we have been exploring proteins derived from common ancestors—that is, through *divergent evolution*. In other cases, clear examples have been found of proteins that are structurally similar in important ways but are not descended from a common ancestor. How might two unrelated proteins come to resemble each other structurally? Two proteins evolving independently may have converged on a similar structure in order to perform a similar biochemical activity. Perhaps that structure was an especially effective solution to a biochemical problem that organisms face. The process by which very different evolutionary pathways lead to the same solution is called *convergent evolution*.

One example of convergent evolution is found among the serine proteases. Cleave peptide bonds by hydrolysis. Figure shows for two such enzymes the structure of the active sites—that is, the sites on the proteins at which the hydrolysis reaction takes place. These active-site structures are remarkably similar. In each case, a serine residue, a histidine residue, and an aspartic acid residue are positioned in space in nearly identical arrangements. As we will see, this is the case because chymotrypsin and subtilisin use the same mechanistic solution to the problem of peptide hydrolysis. At first glance, this similarity might suggest that these proteins are homologous. However, striking differences in the overall structures of these proteins make an evolutionary relationship extremely unlikely. Whereas chymotrypsin consists almost entirely of β sheets, subtilisin contains extensive α-helical structure. Moreover, the key serine, histidine, and aspartic acid residues do not occupy similar positions or even appear in the same order within the two sequences. It is extremely unlikely that two proteins evolving from a common ancestor could have retained similar active-site structures while other aspects of the structure changed so dramatically.

COMPARISON OF RNA SEQUENCES CAN BE A SOURCE OF INSIGHT INTO SECONDARY STRUCTURES

A comparison of homologous RNA sequences can be a source of important insights into evolutionary relationships in a manner similar to that already described. In addition, such comparisons provide clues to the threedimensional structure of the RNA itself. Single-stranded nucleic acid molecules fold back on themselves to form elaborate structures held together by Watson-Crick base-pairing and other interactions. In a family of sequences that form such base-paired structures, base sequences may vary, but base-pairing ability is conserved. Consider, for example, a region from a large RNA molecule present in the ribosomes of all organisms. In the region shown, the *E. coli* sequence has a guanine (G) residue in position 9 and a cytosine (C) residue in position 22, whereas the human sequence has uracil (U) in position 9 and adenine (A) in position 22. Examination of the six sequences reveals that the bases in positions 9 and 22 retain the ability to form a Watson-Crick base pair even though the identities of the bases in these positions vary. Base-pairing ability is also conserved in neighbouring positions; we can deduce that two segments with such compensating mutations are likely to form a double helix. Where sequences are known for several homologous RNA molecules, this type of sequence analysis can often suggest complete secondary structures as well as some additional interactions.

EVOLUTIONARY TREES CAN BE CONSTRUCTED ON THE BASIS OF SEQUENCE INFORMATION

The observation that homology is often manifested as sequence similarity suggests that the evolutionary pathway relating the members of a family of proteins may be deduced by examination of sequence similarity. This approach is based on the notion that sequences that are more similar to one another have had less evolutionary time to diverge from one another than have sequences that are less similar. These sequences can be aligned with the additional constraint that gaps, if present, should be at the same positions in all of the proteins. These aligned sequences can be used to construct an *evolutionary tree* in which the length of the branch connecting each pair of proteins is proportional to the number of amino acid differences between the sequences.

Such comparisons reveal only the relative divergence times—for example, that myoglobin diverged from hemoglobin twice as long ago as the α chain diverged from the β chain. How can we estimate the approximate dates of gene duplications and other evolutionary events? Evolutionary trees can be calibrated by comparing the deduced branch points with divergence times determined from the fossil record. For example, the duplication leading to the two chains of hemoglobin appears to have occurred 350 million years ago. This estimate is supported by the observation that jawless fish such as the lamprey, which diverged from bony fish approximately 400 million years ago, contain hemoglobins built from a single type of subunit.

These methods can be applied to both relatively modern and very ancient molecules, such as the ribosomal RNAs that are found in all organisms. Indeed, it was such an RNA sequence analysis that led to the suggestion that Archaea are a distinct group of organisms that diverged from Bacteria very early in evolutionary history.

MODERN TECHNIQUES MAKE THE EXPERIMENTAL EXPLORATION OF EVOLUTION POSSIBLE

Two techniques of biochemistry have made it possible to examine the course of evolution more directly and not simply by inference. The polymerase chain reaction allows the direct examination of ancient DNA sequences, releasing us, at least in some cases, from the constraints of being able to examine existing genomes from living organisms only. Molecular evolution may be investigated through the use of *combinatorial chemistry,* the process of producing large populations of molecules en masse and selecting for a biochemical property. This exciting process provides a glimpse into the types of molecules that may have existed in the RNA world.

ANCIENT DNA CAN SOMETIMES BE AMPLIFIED AND SEQUENCED

The tremendous chemical stability of DNA makes the molecule well suited to its role as the storage site of genetic information. So stable is the molecule that samples of DNA have survived for many thousands of years under appropriate conditions. With the development of PCR methods, such ancient DNA can sometimes be amplified and sequenced. This approach has been applied to mitochondrial DNA from a Neanderthal fossil estimated at between 30,000 and 100,000 years of age found near Düsseldorf, Germany, in 1856. Investigators managed to identify a total of 379 bases of sequence. Comparison with a number of the corresponding sequences from *Homo sapiens* revealed between 22 and 36 substitutions, considerably fewer than the average of 55 differences between human beings and chimpanzees over the common bases in this region. Further analysis suggested that the common ancestor of modern human beings and Neanderthals lived approximately 600 million years ago. An evolutionary tree constructed by using these and other data revealed that the Neanderthal was not an intermediate between chimpanzees and human beings but, instead, was an evolutionary "dead end" that became extinct.

Note that earlier studies describing the sequencing of much more ancient DNA such as that found in insects trapped in amber appear to have been flawed; contaminating modern DNA was responsible for the sequences determined. Successful sequencing of ancient DNA requires sufficient DNA for reliable amplification and the rigorous exclusion of all sources of contamination.

MOLECULAR EVOLUTION CAN BE EXAMINED EXPERIMENTALLY

Evolution requires three processes:

1. The generation of a diverse population,
2. The selection of members based on some criterion of fitness, and
3. Reproduction to enrich the population in more fit members. Nucleic acid molecules are capable of undergoing all three processes in vitro under appropriate conditions. The results of such studies enable us to glimpse how evolutionary processes might have generated catalytic activities and specific binding abilities—important biochemical functions in all living systems.

A diverse population of nucleic acid molecules can be synthesized in the laboratory by the process of combinatorial chemistry, which rapidly produces large populations of a particular type of molecule such as a nucleic acid. A population of molecules of a given size can be generated randomly so that many or all possible sequences are present in the mixture. When an initial population has been generated, it is subjected to a selection process that isolates specific molecules with desired binding or reactivity properties. Finally, molecules that have survived the selection process are allowed to reproduce through the use of PCR; primers are directed toward specific sequences included at the ends of each member of the population.

As an example of this approach, consider an experiment that set a goal of creating an RNA molecule capable of binding adenosine triphosphate and related nucleotides. Such ATP-bonding molecules are of interest because they might have been present in the RNA world. An initial population of RNA molecules 169 nucleotides long was created; 120 of the positions differed randomly, with equimolar mixtures of adenine, cytosine, guanine, and uracil. The initial synthetic pool that was used contained approximately 10^{14} RNA molecules. Note that this number is a very small fraction of the total possible pool of random 120-base sequences. From this pool, those molecules that bound to ATP, which had been immobilized on a column, were selected.

The collection of molecules that were bound well by the ATP affinity column was allowed to reproduce by reverse transcription into DNA, amplification by PCR, and transcription back into RNA. This new population was subjected to additional rounds of selection for ATP-binding activity. After eight generations, members of the selected population were characterized by sequencing. Seventeen different sequences were obtained. Each of these molecules bound ATP with high affinity, as indicated by dissociation constants less than 50 μM.

The folded structure of the ATP-binding region from one of these RNAs was determined by nuclear magnetic resonance methods. As expected, this 40-nucleotide molecule is composed of two Watson-Crick base-paired helical regions separated by an 11-nucleotide loop. This loop folds back on itself in an intricate way to form a deep pocket into which the adenine ring can fit. Thus, a structure was generated, or evolved, that was capable of a specific interaction.

ENZYMES: BASIC CONCEPTS AND KINETICS

Enzymes, the catalysts of biological systems, are remarkable molecular devices that determine the patterns of chemical transformations. They also mediate the transformation of one form of energy into another. The most striking characteristics of enzymes are their *catalytic power and specificity*. Catalysis takes place at a particular site on the enzyme called the *active site. Nearly all known enzymes are proteins*. However, proteins do not have an absolute monopoly on catalysis; the discovery of catalytically active RNA molecules provides compelling evidence that RNA was an early biocatalyst.

Proteins as a class of macromolecules are highly effective catalysts for an enormous diversity of chemical reactions because of their capacity *to specifically bind a very wide range of molecules*. By utilizing the full repertoire of intermolecular forces, enzymes bring substrates together in an optimal orientation, the prelude to making and breaking chemical bonds. They catalyze reactions *by stabilizing transition states,* the highest-energy species in reaction pathways. By selectively stabilizing a transition state, an enzyme determines which one of several potential chemical reactions actually takes place.

ENZYMES ARE POWERFUL AND HIGHLY SPECIFIC CATALYSTS

Enzymes accelerate reactions by factors of as much as a million or more. Indeed, most reactions in biological systems do not take place at perceptible rates in the absence of enzymes. Even a reaction as simple as the hydration of carbon dioxide is catalyzed by an enzyme—namely, carbonic anhydrase. The transfer of CO_2 from the tissues into the blood and then to the alveolar air would be less complete in the absence of this enzyme. In fact, carbonic anhydrase is one of the fastest enzymes known. Each enzyme molecule can hydrate 10^6 molecules of CO_2 *per second*. This catalyzed reaction is 10^7 times as fast as the uncatalyzed one. We will consider the mechanism of carbonic anhydrase catalysis in Chapter 9. Enzymes are highly specific both in the reactions that they catalyze and in their choice of reactants, which are called *substrates*. An enzyme usually catalyzes a single chemical reaction or a set of closely related reactions. Side reactions leading to the wasteful formation of by-products are rare in enzyme-catalyzed reactions, in contrast with uncatalyzed ones.

$$O{=}C{=}O + H_2O \rightleftharpoons HO{-}\overset{\overset{\large O}{\|}}{C}{-}OH$$

Let us consider *proteolytic enzymes* as an example. In vivo, these enzymes catalyze *proteolysis,* the hydrolysis of a peptide bond.

Most proteolytic enzymes also catalyze a different but related reaction in vitro—namely, the hydrolysis of an ester bond. Such reactions are more easily monitored than is proteolysis and are useful in experimental investigations of these enzymes.

Proteolytic enzymes differ markedly in their degree of substrate specificity. Subtilisin, which is found in certain bacteria, is quite undiscriminating: it will cleave any peptide bond with little regard to the identity of the adjacent side chains. Trypsin, a digestive enzyme, is quite specific and catalyzes the splitting of peptide bonds only on the carboxyl side of lysine and arginine residues. Thrombin, an enzyme that participates in blood clotting, is even more specific than trypsin. It catalyzes the hydrolysis of Arg-Gly bonds in particular peptide sequences only.

DNA polymerase I, a template-directed enzyme is another highly specific catalyst. It adds nucleotides to a DNA strand that is being synthesized, in a sequence determined by the sequence of nucleotides in another DNA strand that serves as a template. DNA polymerase I is remarkably precise in carrying out the instructions given by the template. It inserts the wrong nucleotide into a new DNA strand less than one in a million times.

The specificity of an enzyme is due to the precise interaction of the substrate with the enzyme. This precision is a result of the intricate three-dimensional structure of the enzyme protein.

MANY ENZYMES REQUIRE COFACTORS FOR ACTIVITY

The catalytic activity of many enzymes depends on the presence of small molecules termed *cofactors*, although the precise role varies with the cofactor and the enzyme. Such an enzyme without its cofactor is referred to as an *apoenzyme*; the complete, catalytically active enzyme is called a *holoenzyme*.

Apoenzyme + cofactor = holoenzyme

Cofactors can be subdivided into two groups: metals and small organic molecules. The enzyme carbonic anhydrase, for example, requires Zn^{2+} for its activity. Glycogen phosphorylase which mobilizes glycogen for energy, requires the small organic molecule pyridoxal phosphate (PLP).

Cofactors that are small organic molecules are called *coenzymes*. Often derived from vitamins, coenzymes can be either tightly or loosely bound to the enzyme. If tightly bound, they are called *prosthetic groups*. Loosely associated coenzymes are more like cosubstrates because they bind to and are released from the enzyme just as substrates and products are. The use of the same coenzyme by a variety of enzymes and their source in vitamins sets coenzymes apart from normal substrates, however. Enzymes that use the same coenzyme are usually mechanistically similar.

ENZYMES MAY TRANSFORM ENERGY FROM ONE FORM INTO ANOTHER

In many biochemical reactions, *the energy of the reactants is converted with high efficiency into a different form*. For example, in photosynthesis, light energy is converted into chemical-bond energy through an ion gradient. In mitochondria, the free energy contained in small molecules derived from food is converted first into the free energy of an ion gradient and then into a different currency, the free energy of adenosine triphosphate. Enzymes may then use the chemical-bond energy of ATP in many ways. The enzyme myosin converts the energy of ATP into the mechanical energy of contracting muscles. Pumps in the membranes of cells and organelles, which can be thought of as enzymes that move substrates rather than chemically altering them, create chemical and electrical gradients by using the energy of ATP to transport molecules and ions. The molecular mechanisms of these energy-transducing enzymes are being unraveled. We will see in subsequent chapters how unidirectional cycles of discrete steps—binding, chemical transformation, and release—lead to the conversion of one form of energy into another.

ENZYMES ARE CLASSIFIED ON THE BASIS OF THE TYPES OF REACTIONS THAT THEY CATALYZE

Many enzymes have common names that provide little information about the reactions that they catalyze. For example, a proteolytic enzyme secreted by the pancreas is called trypsin. Most other enzymes are named for their substrates and for the reactions that they catalyze, with the suffix "ase" added. Thus, an ATPase is an enzyme that breaks down ATP, whereas ATP synthase is an enzyme that synthesizes ATP.

To bring some consistency to the classification of enzymes, in 1964 the International Union of Biochemistry established an Enzyme Commission to develop a nomenclature for enzymes. Reactions were divided into six major groups numbered 1 through 6. These groups were subdivided and further subdivided, so that a four-digit number preceded by the letters *EC* for Enzyme Commission could precisely identify all enzymes.

Consider as an example nucleoside monophosphate (NMP) kinase. It catalyzes the following reaction:

$$\text{ATP} + \text{NMP} \rightleftharpoons \text{ADP} + \text{NDP}$$

NMP kinase transfers a phosphoryl group from ATP to NMP to form a nucleoside diphosphate (NDP) and ADP. Consequently, it is a transferase, or member of group 2. Many groups in addition to phosphoryl groups, such as sugars and carbon units, can be transferred. Transferases that shift a phosphoryl group are designated 2.7. Various functional groups can accept the phosphoryl group. If a phosphate is the acceptor, the transferase is designated 2.7.4. The final number designates the acceptor more precisely. In regard to NMP kinase, a nucleoside monophosphate is the acceptor, and the enzyme's designation is EC 2.7.4.4. Although the common names are used routinely, the classification number is used when the precise identity of the enzyme might be ambiguous.

FREE ENERGY IS A USEFUL THERMODYNAMIC FUNCTION FOR UNDERSTANDING ENZYMES

Some of the principles of thermodynamics were introduced in the idea of *free energy (G)*. To fully understand how enzymes operate, we need to consider two thermodynamic properties of the reaction:

1. The free-energy difference ΔG) between the products and reactants and
2. The energy required to initiate the conversion of reactants to products. The former determines whether the reaction will be spontaneous, whereas the later determines the rate of the reaction. Enzymes affect only the latter. First, we will consider the thermodynamics of reactions and then the rates of reactions.

THE FREE-ENERGY CHANGE PROVIDES INFORMATION ABOUT THE SPONTANEITY BUT NOT THE RATE OF A REACTION

As stated in figure the free-energy change of a reaction ΔG) tells us if the reaction can occur spontaneously:

1. *A reaction can occur spontaneously only if* ΔG *is negative.* Such reactions are said to be exergonic.
2. A system is at equilibrium and no net change can take place if ΔG is zero.
3. A reaction cannot occur spontaneously if ΔG is positive. An input of free energy is required to drive such a reaction. These reactions are termed endergonic.

Two additional points need to be emphasized. The ΔG of a reaction depends only on the free energy of the products (the final state) minus the free energy of the reactants (the initial state). *The ΔG of a reaction is independent of the path (or molecular mechanism) of the transformation.* The mechanism of a reaction has no effect on ΔG. For example, the ΔG for the oxidation of glucose to CO_2 and H_2O is the same whether it occurs by combustion in vitro or by a series of enzyme-catalyzed steps in a cell. *The ΔG provides no information about the rate of a reaction.* A negative ΔG indicates that a reaction *can* occur spontaneously, but it does not signify whether it will proceed at a perceptible rate. As will be discussed shortly the rate of a reaction depends on the *free energy of activation* $\Delta G^{\ddagger}$), which is largely unrelated to the ΔG of the reaction.

THE STANDARD FREE-ENERGY CHANGE OF A REACTION IS RELATED TO THE EQUILIBRIUM CONSTANT

As for any reaction, we need to be able to determine ΔG for an enzymecatalyzed reaction in order to know whether the reaction is spontaneous or an input of energy is required. To determine this important thermodynamic parameter, we need to take into account the nature of both the reactants and the products as well as their concentrations.

Consider the reaction

$$A + B \rightleftharpoons C + D$$

The ΔG of this reaction is given by

$$\Delta G = \Delta g^{\circ} + RTIn\frac{[C][D]}{[A][B]} \qquad (1)$$

in which ΔG° is the *standard free-energy change, R* is the gas constant, *T* is the absolute temperature, and [A], [B], [C], and [D] are the molar concentrations (more precisely, the activities) of the reactants. ΔG° is the freeenergy change for this reaction under standard conditions—that is, when each of the reactants A, B, C, and D is present at a concentration of

1.0 M (for a gas, the standard state is usually chosen to be 1 atmosphere). Thus, the ΔG of a reaction depends on the *nature* of the reactants (expressed in the ΔG° term of equation 1) and on their *concentrations* (expressed in the logarithmic term of equation 1).

A convention has been adopted to simplify free-energy calculations for biochemical reactions. The standard state is defined as having a pH of 7. Consequently, when H^+ is a reactant, its activity has the value 1 (corresponding to a pH of 7) in equations 1 and 4 (below). The activity of water also is taken to be 1 in these equations. The *standard free-energy change at pH 7*, denoted by the symbol $\Delta G^{\circ\prime}$ will be used throughout this book. The *kilocalorie* (abbreviated *kcal*) and the *kilojoule (kJ)* will be used as the units of energy. One kilocalorie is equivalent to 4.184 kilojoules.

The relation between the standard free energy and the equilibrium constant of a reaction can be readily derived. This equation is important because it displays the energetic relation between products and reactants in terms of their concentrations. At equilibrium, $\Delta G = 0$. Equation 1 then becomes

$$0 = \Delta G^{\circ\prime} + RT \ln \frac{[C][D]}{[A][B]} \qquad (2)$$

and so

$$\Delta G^{\circ\prime} = -RT \ln \frac{[C][D]}{[A][B]} \qquad (3)$$

The equilibrium constant under standard conditions, K'_{eq}, is defined as

$$K'_{eq} = \frac{[C][D]}{[A][B]} \qquad (4)$$

Substituting equation 4 into equation 3 gives

$$\Delta G^{\circ\prime} = -RT \ln K'_{eq} \qquad (5)$$

$$\Delta G^{\circ\prime} = -2.303RT \log10\ K'eq \tag{6}$$

which can be rearranged to give

$$K'eq = 10^{-\Delta G^{\circ\prime}/(2.303RT)} \tag{7}$$

Substituting $R = 1.987 \times 10^{-3}$ kcal mol^{-1} deg^{-1} and $T = 298$ K (corresponding to 25°C) gives

$$K'eq = 10^{-\Delta G^{\circ\prime}/1.36} \tag{8}$$

where $\Delta G^{\circ\prime}$ is here expressed in kilocalories per mole because of the choice of the units for R in equation 7. Thus, the standard free energy and the equilibrium constant of a reaction are related by a simple expression. For example, an equilibrium constant of 10 gives a standard free-energy change of -1.36 kcal mol^{-1} (–5.69 kJ mol^{-1}) at 25°C. Note that, for each 10-fold change in the equilibrium constant, the ΔG°2 changes by 1.36 kcal mol^{-1} (5.69 kJ mol^{-1}).

As an example, let us calculate $\Delta G^{\circ\prime}$ and ΔG for the isomerization of dihydroxyacetone phosphate (DHAP) to glyceraldehyde 3-phosphate (GAP). This reaction takes place in glycolysis (16.1.4). At equilibrium, the ratio of GAP to DHAP is 0.0475 at 25°C (298 K) and pH 7. Hence, $K2_{eq} = 0.0475$. The standard free-energy change for this reaction is then calculated from equation 6:

$$\begin{aligned}\Delta G^{\circ\prime} &= -2.303RT \log_{10} K'_{eq} \\ &= -2.303 \times 1.987 \times 10^{-3} \times 298 \times \log_{10}(0.0475) \\ &= +1.80 \text{ kcal mol}^{-1} (7.53 \text{ kJ mol}^{-1})\end{aligned}$$

Under these conditions, the reaction is endergonic. DHAP will not spontaneously convert to GAP.

Now let us calculate ΔG for this reaction when the initial concentration of DHAP is 2×10^{-4} M and the initial concentration of GAP is 3×10^{-6} M. Substituting these values into equation 1 gives

$$\begin{aligned}\Delta G &= 1.80 \text{ kcal mol}^{-1} + 2.303RT\log_{10}\frac{3\times10^{-6} M}{2\times10^{-4} M} \\ &= 1.80 \text{ kcal mol}^{-1} - 2.49 \text{ kcal mol}^{-1}\end{aligned}$$

$$= -0.69 \text{ kcal mol}^{-1}(-2.89\text{kJ mol}^{-1})$$

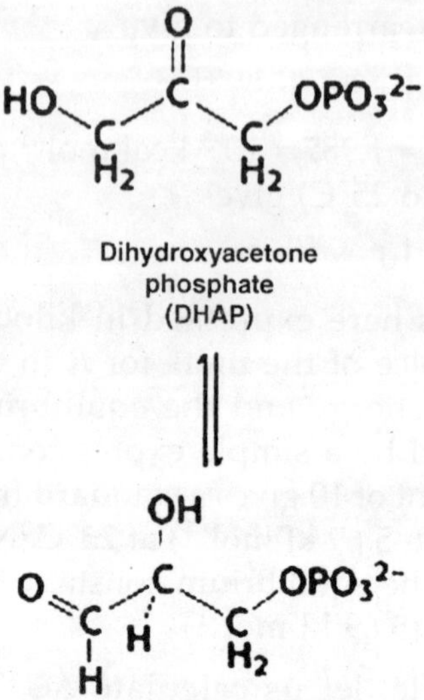

Dihydroxyacetone phosphate (DHAP)

Glyceraldehyde 3-phosphate (GAP)

This negative value for the ΔG indicates that the isomerization of DHAP to GAP is exergonic and can occur spontaneously when these species are present at the aforestated concentrations. Note that ΔG for this reaction is negative, although $\Delta G^{o'}$ is positive. *It is important to stress that whether the ΔG for a reaction is larger, smaller, or the same as $\Delta G^{o'}$ depends on the concentrations of the reactants and products.* The criterion of spontaneity for a reaction is ΔG, not $\Delta G^{o'}$. This point is important because reactions that are not spontaneous based on $\Delta G^{o'}$ can be made spontaneous by adjusting the concentrations of reactants and products. This principle is the basis of the coupling of reactions to form metabolic pathways.

ENZYMES ALTER ONLY THE REACTION RATE AND NOT THE REACTION EQUILIBRIUM

Because enzymes are such superb catalysts, it is tempting to ascribe to them powers that they do not have. An enzyme cannot alter the laws of thermodynamics and *consequently cannot alter the equilibrium of a chemical reaction.* This inability means that an enzyme accelerates the forward and reverse reactions by precisely the same factor. Consider the interconversion of A and B. Suppose that, in the absence of enzyme, the forward rate constant (k_F) is 10^{-4} s^{-1} and the reverse rate constant (k_R) is 10^{-6} s^{-1}. The equilibrium constant K is given by the ratio of these rate constants:

$$A \underset{10^{-6}\,s^{-1}}{\overset{10^{-4}\,s^{-1}}{\rightleftharpoons}} B$$

$$K = \frac{[B]}{[A]} = \frac{k_F}{k_R} = \frac{10^{-4}}{10^{-6}} = 100$$

The equilibrium concentration of B is 100 times that of A, whether or not enzyme is present. However, it might take considerable time to approach this equilibrium without enzyme, whereas equilibrium would be attained rapidly in the presence of a suitable enzyme. *Enzymes accelerate the attainment of equilibria but do not shift their positions. The equilibrium position is a function only of the free-energy difference between reactants and products.*

ENZYMES ACCELERATE REACTIONS BY FACILITATING THE FORMATION OF THE TRANSITION STATE

The free-energy difference between reactants and products accounts for the equilibrium of the reaction, but enzymes accelerate how quickly this equilibrium is attained. How can we explain the rate enhancement in terms of thermodynamics? To do so, we have to consider not the end points of the reaction but the chemical pathway between the end points.

A chemical reaction of substrate S to form product P goes through a *transition state* $S^{\ddagger}$ that has a higher free energy than does either S or P. The double dagger denotes a thermodynamic property of the transition state. The transition state is the most seldom occupied species along the reaction pathway because it is the one with the highest free energy. The difference in free energy between the transition state and the substrate is called the *Gibbs free energy of activation* or simply the *activation energy*, symbolized by $\Delta G^{\ddagger}$

$$\Delta G' = GS^{\ddagger} - G_S$$

Note that the energy of activation, or $\Delta G^{\ddagger}$, does not enter into the final ΔG calculation for the reaction, because the energy input required to reach the transition state is returned when the transition state forms the product. The activation-energy barrier immediately suggests how enzymes enhance reaction rate without altering ΔG of the reaction: enzymes function to lower the activation energy, or, in other words, *enzymes facilitate the formation of the transition state.*

One approach to understanding how enzymes achieve this facilitation is to assume that the transition state ($S^{\ddagger}$) and the substrate (S) are in equilibrium.

$$S \underset{}{\overset{K^{\ddagger}}{\rightleftharpoons}} S^{\ddagger} \overset{v}{\longrightarrow} P$$

In which $K^{\ddagger}$ is the equilibrium constant for the formation of $S^{\ddagger}$, and v is the rate of formation of product from $S^{\ddagger}$.

The rate of the reaction is proportional to the concentration of $S^{\ddagger}$: because only $S^{\ddagger}$

Rate α $[S^{\ddagger}]$

can be converted into product. The concentration of $S^{\ddagger}$ is in turn related to the free energy difference between $S^{\ddagger}$ and S, because these two chemical species are assumed to be in equilibrium; the greater the difference between these two states, the smaller the amount of $S^{\ddagger}$.

Because the reaction rate is proportional to the concentration of $S^{\ddagger}$, and the concentration of $S^{\ddagger}$ depends on $DG^{\ddagger}$, the rate of reaction V depends on $DG^{\ddagger}$. Specifically,

$$V = v[S^{\ddagger}] = \frac{kT}{h}[S]e^{-\Delta G^{\ddagger}/RT}$$

In this equation, k is Boltzmann's constant, and h is Planck's constant. The value of kT/h at 25°C is 6.2×10^{12} s^{-1}. Suppose that the free energy of activation is 6.82 kcal mol^{-1}

(28.53 kJ mol^{-1}). The ratio [S‡]/[S] is then 10^{-5}. If we assume for simplicity's sake that [S] = 1 M, then the reaction rate *V* is 6.2 × 10^7 s^{-1}. If ΔG‡ were lowered by 1.36 kcal mol^{-1} (5.69 kJ mol^{-1}), the ratio [S‡]/[S] is then 10^{-4}, and the reaction rate would be 6.2 × 10^8 s^{-1}.

Thus, we see the key to how enzymes operate: *Enzymes accelerate reactions by decreasing ΔG‡, the activation energy.* The combination of substrate and enzyme creates a new reaction pathway whose transition-state energy is lower than that of the reaction in the absence of enzyme. The lower activation energy means that more molecules have the required energy to reach the transition state. Decreasing the activation barrier is analogous to lowering the height of a high-jump bar; more athletes will be able to clear the bar. *The essence of catalysis is specific binding of the transition state.*

THE FORMATION OF AN ENZYME-SUBSTRATE COMPLEX IS THE FIRST STEP IN ENZYMATIC CATALYSIS

Much of the catalytic power of enzymes comes from their bringing substrates together in favourable orientations to promote the formation of the transition states in *enzyme-substrate* (ES) complexes. The substrates are bound to a specific region of the enzyme called the *active site.* Most enzymes are highly selective in the substrates that they bind. Indeed, the catalytic specificity of enzymes depends in part on the specificity of binding.

What is the evidence for the existence of an enzyme-substrate complex?

1. The first clue was the observation that, at a constant concentration of enzyme, the reaction rate increases with increasing substrate concentration until a maximal velocity is reached. In contrast, uncatalyzed reactions do not show this saturation effect. *The fact that an enzyme-catalyzed reaction has a maximal velocity suggests the formation of a discrete ES complex.* At a

sufficiently high substrate concentration, all the catalytic sites are filled and so the reaction rate cannot increase. Although indirect, this is the most general evidence for the existence of ES complexes.

2. *X-ray crystallography* has provided high-resolution images of substrates and substrate analogs bound to the active sites of many enzymes. X-ray studies carried out at low temperatures (to slow reactions down) are providing revealing views of enzyme-substrate complexes and their subsequent reactions. A new technique, *time-resolved crystallography,* depends on cocrystallizing a photolabile substrate analog with the enzyme. The substrate analog can be converted to substrate light, and images of the enzyme- substrate complex are obtained in a fraction of a second by scanning the crystal with intense, polychromatic x-rays from a synchrotron.

3. The *spectroscopic characteristics* of many enzymes and substrates change on formation of an ES complex. These changes are particularly striking if the enzyme contains a coloured prosthetic group. Tryptophan synthetase, a bacterial enzyme that contains a pyridoxal phosphate (PLP) prosthetic group, provides a nice illustration. This enzyme catalyzes the synthesis of l-tryptophan from l-serine and indole-derivative. The addition of l-serine to the enzyme produces a marked increase in the fluorescence of the PLP group. The subsequent addition of indole, the second substrate, reduces this fluorescence to a level even lower than that of the enzyme alone. Thus, fluorescence spectroscopy reveals the existence of an enzyme-serine complex and of an enzyme-serine-indole complex. Other spectroscopic techniques, such as nuclear magnetic resonance and electron spin resonance, also are highly informative about ES interactions.

THE ACTIVE SITES OF ENZYMES HAVE SOME COMMON FEATURES

The active site of an enzyme is the region that binds the substrates (and the cofactor, if any). It also contains the residues that directly participate in the making and breaking of bonds. These residues are called the *catalytic groups*. In essence, *the interaction of the enzyme and substrate at the active site promotes the formation of the transition state*. The active site is the region of the enzyme that most directly lowers the $\Delta G^{\ddagger}$ of the reaction, which results in the rate enhancement characteristic of enzyme action. Although enzymes differ widely in structure, specificity, and mode of catalysis, a number of generalizations concerning their active sites can be stated:

1. *The active site is a three-dimensional cleft formed by groups that come from different parts of the amino acid sequence*—indeed, residues far apart in the sequence may interact more strongly than adjacent residues in the amino acid sequence. In lysozyme, an enzyme that degrades the cell walls of some bacteria, the important groups in the active site are contributed by residues numbered 35, 52, 62, 63, 101, and 108 in the sequence of the 129 amino acids.
2. *The active site takes up a relatively small part of the total volume of an enzyme*. Most of the amino acid residues in an enzyme are not in contact with the substrate, which raises the intriguing question of why enzymes are so big. Nearly all enzymes are made up of more than 100 amino acid residues, which gives them a mass greater than 10 kd and a diameter of more than 25 Å. The "extra" amino acids serve as a scaffold to create the three-dimensional active site from amino acids that are far apart in the primary structure. Amino acids near to one another in the primary structure are often sterically constrained from adopting the structural relations necessary to form the active site. In many proteins, the remaining amino acids also constitute regulatory sites, sites of

interaction with other proteins, or channels to bring the substrates to the active sites.

3. *Active sites are clefts or crevices.* In all enzymes of known structure, substrate molecules are bound to a cleft or crevice. Water is usually excluded unless it is a reactant. The nonpolar character of much of the cleft enhances the binding of substrate as well as catalysis. Nevertheless, the cleft may also contain polar residues. In the nonpolar microenvironment of the active site, certain of these polar residues acquire special properties essential for substrate binding or catalysis. The internal positions of these polar residues are biologically crucial exceptions to the general rule that polar residues are exposed to water.
4. *Substrates are bound to enzymes by multiple weak attractions.* ES complexes usually have equilibrium constants that range from 10^{-2} to 10^{-8} M, corresponding to free energies of interaction ranging from about -3 to -12 kcal mol^{-1} (from -13 to -50 kJ mol^{-1}). The noncovalent interactions in ES complexes are much weaker than covalent bonds, which have energies between -50 and -110 kcal mol^{-1} (between -210 and -460 kJ mol^{-1}). Van der Waals forces become significant in binding only when numerous substrate atoms simultaneously come close to many enzyme atoms. Hence, the enzyme and substrate should have complementary shapes. The directional character of hydrogen bonds between enzyme and substrate often enforces a high degree of specificity, as seen in the RNA-degrading enzyme ribonuclease.
5. *The specificity of binding depends on the precisely defined arrangement of atoms in an active site.* Because the enzyme and the substrate interact by means of short-range forces that require close contact, a substrate must have a matching shape to fit into the site. Emil Fischer's analogy of the lock and key expressed in 1890, has proved to be highly stimulating and

fruitful. However, we now know that enzymes are flexible and that the shapes of the active sites can be markedly modified by the binding of substrate, as was postulated by Daniel E. Koshland, Jr., in 1958. The active sites of some enzymes assume a shape that is complementary to that of the transition state only *after* the substrate is bound. This process of dynamic recognition is called *induced fit*.

THE MICHAELIS-MENTEN MODEL ACCOUNTS FOR THE KINETIC PROPERTIES OF MANY ENZYMES

The primary function of enzymes is to enhance rates of reactions so that they are compatible with the needs of the organism. To understand how enzymes function, we need a kinetic description of their activity. For many enzymes, the rate of catalysis V_0, which is defined as the number of moles of product formed per second, varies with the substrate concentration [S] in a manner shown in figure. The rate of catalysis rises linearly as substrate concentration increases and then begins to level off and approach a maximum at higher substrate concentrations. Before we can accurately interpret this graph, we need to understand how it is generated. Consider an enzyme that catalyzes the S to P by the following pathway:

$$E + S \underset{k_{-1}}{\overset{k_1}{\rightleftharpoons}} ES \underset{k_{-2}}{\overset{k_2}{\rightleftharpoons}} E + P$$

The extent of product formation is determined as a function of time for a series of substrate concentrations. As expected, in each case, the amount of product formed increases with time, although eventually a time is reached when there is *no net change* in the concentration of S or P. The enzyme is still actively converting substrate into product and visa versa, but the reaction equilibrium has been attained.

Enzyme kinetics are more easily approached if we can ignore the back reaction. We define V_0 as the rate of increase in product with time when [P] is low; that is, at times close to zero (hence, V_0). Thus, V_0 is determined for each substrate concentration by measuring the rate of product formation at early times before P accumulates.

At a fixed concentration of enzyme, V_0 is almost linearly proportional to [S] when [S] is small but is nearly independent of [S] when [S] is large. In 1913, Leonor Michaelis and Maud Menten proposed a simple model to account for these kinetic characteristics. The critical feature in their treatment is that a specific ES complex is a necessary intermediate in catalysis. The model proposed, which is the simplest one that accounts for the kinetic properties of many enzymes, is

$$E + S \underset{k}{\overset{k_1}{\rightleftharpoons}} ES \xrightarrow{K_2} E + P \tag{9}$$

An enzyme E combines with substrate S to form an ES complex, with a rate constant k_1. The ES complex has two possible fates. It can dissociate to E and S, with a rate constant k_{-1}, or it can proceed to form product P, with a rate constant k_2. Again, we assume that almost none of the product reverts to the initial substrate, a condition that holds in the initial stage of a reaction before the concentration of product is appreciable.

We want an expression that relates the rate of catalysis to the concentrations of substrate and enzyme and the rates of the individual steps. Our starting point is that the catalytic rate is equal to the product of the concentration of the ES complex and k_2.

$$V_0 = k_2[ES] \tag{10}$$

Now we need to express [ES] in terms of known quantities. The rates of formation and breakdown of ES are given by:

$$\text{Rate of formation of ES} = K_1[E][S] \tag{11}$$

$$\text{Rate of breakdown of ES} = (k_{-1} + k_2)[ES] \tag{12}$$

To simplify matters, we will work under the *steady-state assumption*. In a steady state, the concentrations of intermediates, in this case [ES], stay the same even if the concentrations of starting materials and products are changing.

This occurs when the rates of formation and breakdown of the ES complex are equal. Setting the right-hand sides of equations 11 and 12 equal gives

$$K_1[E][S] = (k_{-1} + k_2)[ES] \tag{13}$$

By rearranging equation 13, we obtain

$$[E][S]/[ES] = (k_{-1} + k_2)/k_1 \tag{14}$$

Equation 14 can be simplified by defining a new constant, K_M, called the *Michaelis constant:*

$$k_M = \frac{k_{-1} + k_2}{k_1} \tag{15}$$

Note that K_M has the units of concentration. K_M is an important characteristic of enzyme-substrate interactions and is independent of enzyme and substrate concentrations.

Inserting equation 15 into equation 14 and solving for [ES] yields

$$[ES] = \frac{[E][S]}{K_M} \tag{16}$$

Now let us examine the numerator of equation 16. The concentration of uncombined substrate [S] is very nearly equal to the total substrate concentration, provided that the concentration of enzyme is much lower than that of substrate. The concentration of uncombined enzyme [E] is equal to the total enzyme concentration $[E]_T$ minus the concentration of the ES complex.

$$[E] = [E]_T - [ES] \tag{17}$$

Substituting this expression for [E] in equation 16 gives

$$[ES] = \frac{([E]_T - [ES])[S]}{K_M} \tag{18}$$

Solving equation 18 for [ES] gives

$$[ES] = \frac{[E]_T [S]/K_M}{1 + [S]/K_M} \tag{19}$$

or

$$[ES] = [E]_T \frac{[S]}{[S]+K_M} \tag{20}$$

By substituting this expression for [ES] into equation 10, we obtain

$$V_0 = k_2[E]_T \frac{[S]}{[S]+K_M} \tag{21}$$

The maximal rate, V_{max}, is attained when the catalytic sites on the enzyme are saturated with substrate—that is, when [ES] = $[E]_T$. Thus,

$$V_{max} = k_2[E]T \tag{22}$$

Substituting equation 22 into equation 21 yields the *Michaelis-Menten equation:*

$$V_0 = V_{max} \frac{[S]}{[S]+K_M} \tag{23}$$

At very low substrate concentration, when [S] is much less than K_M, $V_0 = (V_{max}/K_M)[S]$; that is, the rate is directly proportional to the substrate concentration. At high substrate concentration, when [S] is much greater than K_M, $V_0 = V_{max}$; that is, the rate is maximal, independent of substrate concentration.

The meaning of K_M is evident from equation 23. When $[S] = K_M$, then $V_0 = V_{max}/2$. Thus, K_M *is equal to the substrate concentration at which the reaction rate is half its maximal value.* K_M is an important characteristic of an enzyme-catalyzed reaction and is significant for its biological function.

The physiological consequence of K_M is illustrated by the sensitivity of some individuals to ethanol. Such persons exhibit facial flushing and rapid heart rate (tachycardia) after ingesting even small amounts of alcohol. In the liver, alcohol dehydrogenase converts ethanol into acetaldehyde.

$$CH_3CH_2OH + NAD^+ \xrightleftharpoons{\text{Alchohol dehydrogenase}} CH_3CH0 + H^- + NADH$$

Normally, the acetaldehyde, which is the cause of the symptoms when present at high concentrations, is processed to acetate by acetaldehyde dehydrogenase.

$$CH_3CHO + NAD^+ \xrightleftharpoons{\text{Alchohol dehydrogenase}} CH_3COO^- + NADH + 2H^-$$

Most people have two forms of the acetaldehyde dehydrogenase, a low K_M mitochondrial form and a high K_M cytosolic form. In susceptible persons, the mitochondrial enzyme is less active due to the substitution of a single amino acid, and acetaldehyde is processed only by the cytosolic enzyme. Because this enzyme has a high K_M, less acetaldehyde is converted into acetate; excess acetaldehyde escapes into the blood and accounts for the physiological effects.

THE SIGNIFICANCE OF K_M AND V_{MAX} VALUES

The Michaelis constant, K_M, and the maximal rate, V_{max}, can be readily derived from rates of catalysis measured at a variety of substrate concentrations if an enzyme operates according to the simple scheme given in equation 23. The derivation of K_M and V_{max} is most commonly achieved with the use of curve-fitting programmes on a computer. The K_M values of enzymes range widely. For most enzymes, K_M lies between 10^{-1} and 10^{-7} M. The K_M value for an enzyme depends on the particular substrate and on environmental conditions such as pH, temperature, and ionic strength. The Michaelis constant, K_M, has two meanings. First, K_M is the concentration of substrate at which half the active sites are filled. Thus, K_M provides a measure of the substrate concentration required for significant catalysis to occur. In fact, for many enzymes, experimental evidence suggests that K_M provides an approximation of substrate concentration in vivo. When the K_M is known, the fraction of sites filled, f_{ES}, at any substrate concentration can be calculated from

$$f_{ES} = \frac{V}{V_{max}} = \frac{[S]}{[S] + K_M} \qquad (24)$$

Second, K_M is related to the rate constants of the individual steps in the catalytic scheme given in equation 9. In equation 15, K_M is defined as $(k_{-1} + k_2)/k_1$. Consider a limiting case in which k_{-1} is much greater than k_2. Under such circumstances, the ES complex dissociates to E and S much more rapidly than product is formed. Under these conditions ($k_{-1} >> k_2$),

$$K_M = \frac{k-1}{k_1} \tag{25}$$

The dissociation constant of the ES complex is given by

$$K_{ES} = \frac{[E][S]}{[ES]} = \frac{k_{-1}}{k_1} \tag{26}$$

In other words, K_M *is equal to the dissociation constant of the ES complex if* k_2 *is much smaller than* k_{-1}. When this condition is met, K_M is a measure of the strength of the ES complex: a high K_M indicates weak binding; a low K_M indicates strong binding. It must be stressed that K_M indicates the affinity of the ES complex only when k_{-1} is much greater than k_2.

The maximal rate, V_{max}, reveals the *turnover number* of an enzyme, which is *the number of substrate molecules converted into product by an enzyme molecule in a unit time when the enzyme is fully saturated with substrate.* It is equal to the kinetic constant k_2, which is also called k_{cat}. The maximal rate, V_{max}, reveals the turnover number of an enzyme if the concentration of active sites $[E]_T$ is known, because

$$V_{max} = k_2[E]T$$

and thus

$$k_2 = V_{max}/[E]_T \tag{27}$$

For example, a 10^{-6} M solution of carbonic anhydrase catalyzes the formation of 0.6 M H_2CO_3 per second when it is fully saturated with substrate. Hence, k_2 is 6×10^5 s^{-1}. This turnover number is one of the largest known. Each catalyzed reaction takes place in a time equal to $1/k_2$, which is 1.7 µs for carbonic anhydrase. The turnover numbers of most enzymes with their physiological substrates fall in the range from 1 to 10^4 per second.

KINETIC PERFECTION IN ENZYMATIC CATALYSIS: THE K_{CAT}/K_M CRITERION

When the substrate concentration is much greater than K_M, the rate of catalysis is equal to k_{cat}, the turnover number. However, most enzymes are not normally saturated with substrate. Under physiological conditions, the $[S]/K_M$ ratio is typically between 0.01 and 1.0. When $[S] \ll K_M$, the enzymatic rate is much less than k_{cat} because most of the active sites are unoccupied. Is there a number that characterizes the kinetics of an enzyme under these more typical cellular conditions? Indeed there is, as can be shown by combining equations 10 and 16 to give

$$V_0 = \frac{k_{cat}}{K_M}[E][S] \qquad (28)$$

When $[S] \ll K_M$, the concentration of free enzyme, [E], is nearly equal to the total concentration of enzyme $[E_T]$, so

$$V_0 = \frac{k_{cat}}{K_M}[S][E]_T \qquad (29)$$

Thus, when $[S] \ll K_M$, the enzymatic velocity depends on the values of k_{cat}/K_M, [S], and $[E]_T$. Under these conditions, k_{cat}/K_M is the rate constant for the interaction of S and E and can be used as a measure of catalytic efficiency. For instance, by using k_{cat}/K_M values, one can compare an enzyme's preference for different substrates. Chymotrypsin clearly has a preference for cleaving next to bulky, hydrophobic side chains.

How efficient can an enzyme be? We can approach this question by determining whether there are any physical limits on the value of k_{cat}/K_M. Note that this ratio depends on k_1, k_{-1}, and k_{cat}, as can be shown by substituting for K_M.

$$\frac{k_{cat}}{K_M} = \frac{k_{cat}}{(k_{-1} + k_{cat})/k_1} = \frac{k_{cat}}{k_{cat} + k_{-1}} k_1 < k_1 \qquad (30)$$

Suppose that the rate of formation of product (k_{cat}) is much faster than the rate of dissociation of the ES complex (k_{-1}). The value of k_{cat}/K_M then approaches k_1. Thus, the ultimate limit on the value of k_{cat}/K_M is set by k_1, the rate of formation of the ES complex. *This rate cannot be faster than the diffusion-controlled encounter of an enzyme and its substrate.* Diffusion limits the value of k_1 so that it cannot be higher than between 10^8 and 10^9 s^{-1} M^{-1}. Hence, the upper limit on k_{cat}/K_M is between 10^8 and 10^9 s^{-1} M^{-1}.

The k_{cat}/K_M ratios of the enzymes superoxide dismutase, acetylcholinesterase, and triose phosphate isomerase are between 10^8 and 10^9 s^{-1} M^{-1}. Enzymes such as these that have k_{cat}/K_M ratios at the upper limits have attained *kinetic perfection. Their catalytic velocity is restricted only by the rate at which they encounter substrate in the solution.* Any further gain in catalytic rate can come only by decreasing the time for diffusion. Remember that the active site is only a small part of the total enzyme structure. Yet, for catalytically perfect enzymes, every encounter between enzyme and substrate is productive. In these cases, there may be attractive electrostatic forces on the enzyme that entice the substrate to the active site. These forces are sometimes referred to poetically as *Circe effects*.

CIRCE EFFECT

The utilization of attractive forces to lure a substrate into a site in which it undergoes a transformation of structure, as defined by William P. Jencks, an enzymologist, who coined the term.

A goddess of Greek mythology, Circe lured Odysseus's men to her house and then transformed them into pigs.

The limit imposed by the rate of diffusion in solution can also be partly overcome by confining substrates and products in the limited volume of a multienzyme complex. Indeed, some series of enzymes are associated into organized assemblies so that the product of one enzyme is very rapidly found by the next enzyme. In effect, products are channeled from one enzyme to the next, much as in an assembly line.

MOST BIOCHEMICAL REACTIONS INCLUDE MULTIPLE SUBSTRATES

Most reactions in biological systems usually include two substrates and two products and can be represented by the bisubstrate reaction:

$$A + B \rightleftharpoons P + Q$$

The majority of such reactions entail the transfer of a functional group, such as a phosphoryl or an ammonium group, from one substrate to the other. In oxidation-reduction reactions, electrons are transferred between substrates. Multiple substrate reactions can be divided into two classes: *sequential displacement* and *double displacement*.

SEQUENTIAL DISPLACEMENT

In the sequential mechanism, all substrates must bind to the enzyme before any product is released. Consequently, in a bisubstrate reaction, a *ternary complex* of the enzyme and both substrates forms. Sequential mechanisms are of two types: ordered, in which the substrates bind the enzyme in a defined sequence, and random.

Many enzymes that have NAD^+ or NADH as a substrate exhibit the sequential ordered mechanism. Consider lactate dehydrogenase, an important enzyme in glucose metabolism. This enzyme reduces pyruvate to lactate while oxidizing NADH to NAD^+.

$$\underset{\text{Pyruvate}}{CH_3COCOO^-} + NADH + H^+ \rightleftharpoons \underset{\text{Lactate}}{CH_3CH(OH)COO^-} + NAD^+$$

In the ordered sequential mechanism, the coenzyme always binds first and the lactate is always released first. This sequence can be represented as follows in a notation developed by W. Wallace Cleland:

The enzyme exists as a ternary complex: first, consisting of the enzyme and substrates and, after catalysis, the enzyme and products.

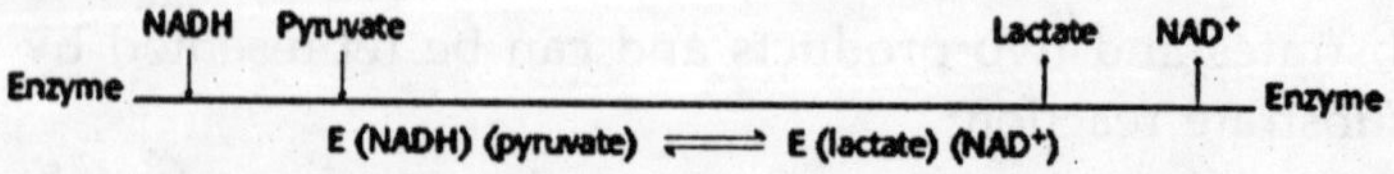

Enzyme — NADH, Pyruvate — E (NADH) (pyruvate) ⇌ E (lactate) (NAD^+) — Lactate, NAD^+ — Enzyme

In the random sequential mechanism, the order of addition of substrates and release of products is random. Sequential random reactions are illustrated by the formation of phosphocreatine and ADP from ATP and creatine, a reaction catalyzed by creatine kinase.

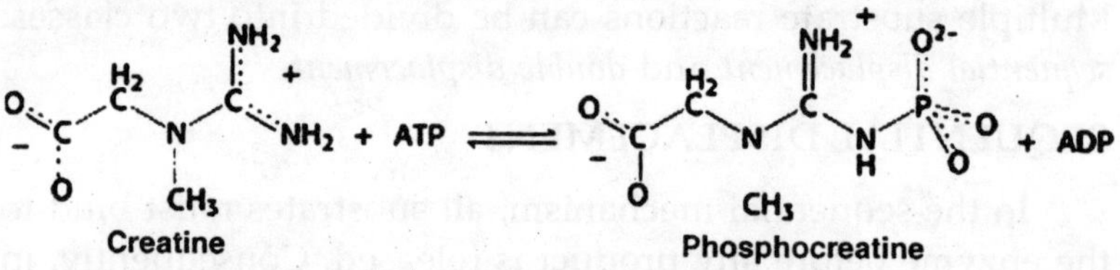

Creatine + ATP ⇌ Phosphocreatine + ADP

Phosphocreatine is an important energy source in muscle. Sequential random reactions can also be depicted in the Cleland notation.

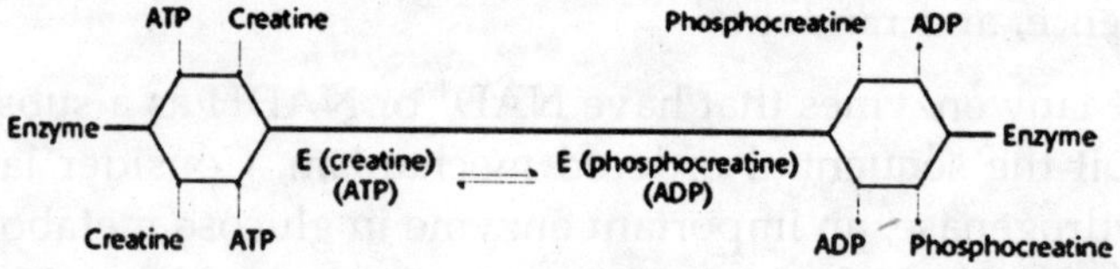

Enzyme — ATP, Creatine / Creatine, ATP — E (creatine) (ATP) ⇌ E (phosphocreatine) (ADP) — Phosphocreatine, ADP / ADP, Phosphocreatine — Enzyme

Although the order of certain events is random, the reaction still passes through the ternary complexes including, first, substrates and, then, products.

DOUBLE-DISPLACEMENT (PING-PONG) REACTIONS

In double-displacement, or Ping-Pong, reactions, one or more products are released before all substrates bind the enzyme. The defining feature of double-displacement reactions is the existence of a *substituted enzyme intermediate,* in which the enzyme is temporarily modified. Reactions that shuttle amino groups between amino acids and α-keto acids are classic examples of double-displacement mechanisms. The enzyme aspartate aminotransferase catalyzes the transfer of an amino group from aspartate to α-ketoglutarate.

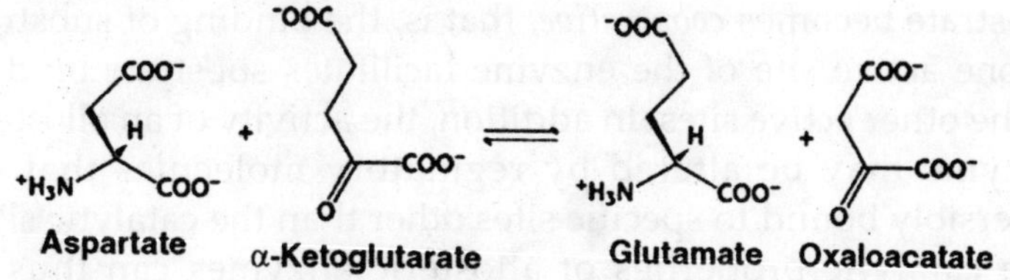

Aspartate + α-Ketoglutarate ⇌ Glutamate + Oxaloacatate

The sequence of events can be portrayed as the following diagram.

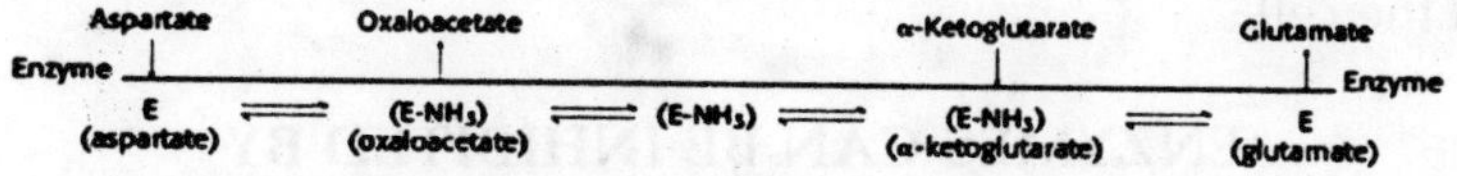

Enzyme — Aspartate — Oxaloacetate — α-Ketoglutarate — Glutamate — Enzyme

E (aspartate) ⇌ (E-NH_3) (oxaloacetate) ⇌ (E-NH_3) ⇌ (E-NH_3) (α-ketoglutarate) ⇌ E (glutamate)

After aspartate binds to the enzyme, the enzyme removes aspartate's amino group to form the substituted enzyme intermediate. The first product, oxaloacetate, subsequently departs. The second substrate, α-ketoglutarate, binds to the enzyme, accepts the amino group from the modified enzyme, and is then released as the final product, glutamate. In the Cleland notation, the substrates appear to bounce on and off the enzyme analogously to a Ping-Pong ball bouncing on a table.

ALLOSTERIC ENZYMES DO NOT OBEY MICHAELIS-MENTEN KINETICS

The Michaelis-Menten model has greatly assisted the development of enzyme chemistry. Its virtues are simplicity and broad applicability. However, the Michaelis-Menten model cannot account for the kinetic properties of many enzymes. An important group of enzymes that do not obey Michaelis-Menten kinetics comprises the *allosteric enzymes*. These enzymes consist of multiple subunits and multiple active sites.

Allosteric enzymes often display sigmoidal plot of the reaction velocity V_0 versus substrate concentration [S], rather than the hyperbolic plots predicted by the Michaelis-Menten equation (equation 23). In allosteric enzymes, the binding of substrate to one active site can affect the properties of other active sites in the same enzyme molecule. A possible outcome of this interaction between subunits is that the binding of substrate becomes *cooperative;* that is, the binding of substrate to one active site of the enzyme facilitates substrate binding to the other active sites. In addition, the activity of an allosteric enzyme may be altered by regulatory molecules that are reversibly bound to specific sites other than the catalytic sites. The catalytic properties of allosteric enzymes can thus be adjusted to meet the immediate needs of a cell. For this reason, allosteric enzymes are key regulators of metabolic pathways in the cell.

ENZYMES CAN BE INHIBITED BY SPECIFIC MOLECULES

The activity of many enzymes can be inhibited by the binding of specific small molecules and ions. This means of inhibiting enzyme activity serves as a major control mechanism in biological systems. The regulation of allosteric enzymes typifies this type of control. In addition, many drugs and toxic agents act by inhibiting enzymes. Inhibition by particular chemicals can be a source of insight into the mechanism of enzyme action: specific inhibitors can often be used to identify residues critical for catalysis. The value of transition-state analogs as potent inhibitors will be discussed shortly.

Enzyme inhibition can be either reversible or irreversible. An *irreversible inhibitor* dissociates very slowly from its target enzyme because it has become tightly bound to the enzyme, either covalently or noncovalently. Some irreversible inhibitors are important drugs. Penicillin acts by covalently modifying the enzyme transpeptidase, thereby preventing the synthesis of bacterial cell walls and thus killing the bacteria. Aspirin acts by covalently modifying the enzyme cyclooxygenase, reducing the synthesis of inflammatory signals.

Reversible inhibition, in contrast with irreversible inhibition, is characterized by a rapid dissociation of the enzyme-inhibitor complex. In *competitive inhibition,* an enzyme can bind substrate (forming an ES complex) or inhibitor (EI) but not both (ESI). The competitive inhibitor resembles the substrate and binds to the active site of the enzyme. The substrate is thereby prevented from binding to the same active site. *A competitive inhibitor diminishes the rate of catalysis by reducing the proportion of enzyme molecules bound to a substrate.* At any given inhibitor concentration, competitive inhibition can be relieved by increasing the substrate concentration. Under these conditions, the substrate "outcompetes" the inhibitor for the active site. Methotrexate is a structural analog of tetrahydrofolate, a coenzyme for the enzyme dihydrofolate reductase, which plays a role in the biosynthesis of purines and pyrimidines. It binds to dihydrofolate reductase 1000-fold more tightly than the natural substrate and inhibits nucleotide base synthesis. It is used to treat cancer.

In *noncompetitive inhibition,* which also is reversible, the inhibitor and substrate can bind simultaneously to an enzyme molecule at different binding sites. A noncompetitive inhibitor acts by decreasing the turnover number rather than by diminishing the proportion of enzyme molecules that are bound to substrate. Noncompetitive inhibition, in contrast with competitive inhibition, cannot be overcome by increasing the substrate concentration. A more complex pattern, called *mixed inhibition,* is produced when a single inhibitor both hinders the binding of substrate and decreases the turnover number of the enzyme.

COMPETITIVE AND NONCOMPETITIVE INHIBITION ARE KINETICALLY DISTINGUISHABLE

How can we determine whether a reversible inhibitor acts by competitive or noncompetitive inhibition? Let us consider only enzymes that exhibit Michaelis-Menten kinetics. Measurements of the rates of catalysis at different concentrations of substrate and inhibitor serve to distinguish the two types of inhibition. In *competitive inhibition,* the inhibitor competes with the substrate for the active site. The dissociation constant for the inhibitor is given by.

$$K_i = [E][I]/[EI]$$

Because increasing the amount of substrate can overcome the inhibition, V_{max} can be attained in the presence of a competitive inhibitor. *The hallmark of competitive inhibition is that it can be overcome by a sufficiently high concentration of substrate.* However, the apparent value of K_M is altered; the effect of a competitive inhibitor is to increase the apparent value of K_M. This new value of K_M, called K^{app}_M, is numerically equal to

$$K_M^{app} = K_M^{(1+[I]/K_i)}$$

where [I] is the concentration of inhibitor and K_i is the dissociation constant for the enzyme-inhibitor complex. As the value of [I] increases, the value of K^{app}_M increases. In the presence of a competitive inhibitor, an enzyme will have the same V_{max} as in the absence of an inhibitor.

In *noncompetitive inhibition,* substrate can still bind to the enzyme-inhibitor complex. However, the enzyme-inhibitor-substrate complex *does not* proceed to form product. The value of V_{max} is decreased to a new value called V^{app}_{max} while the value of K_M is unchanged. Why is V_{max} lowered while K_M remains unchanged? In essence, the inhibitor simply lowers the concentration of functional enzyme. The remaining enzyme behaves like a more dilute solution of enzyme; V_{max} is lower, but K_M is the same. *Noncompetitive inhibition cannot be overcome by increasing the substrate concentration.*

IRREVERSIBLE INHIBITORS CAN BE USED TO MAP THE ACTIVE SITE

The first step in obtaining the chemical mechanism of an enzyme is to determine what functional groups are required for enzyme activity. How can we ascertain these functional groups? X-ray crystallography of the enzyme bound to its substrate provides one approach. Irreversible inhibitors that covalently bond to the enzyme provide an alternative and often complementary means for elucidating functional groups at the enzyme active site because they modify the functional groups, which can then be identified. Irreversible inhibitors can be divided into three categories: group-specific reagents, substrate analogs, and suicide inhibitors.

Group-specific reagents react with specific R groups of amino acids. Two examples of group-specific reagents are diisopropylphosphofluoridate and iodoacetamide. DIPF modifies only 1 of the 28 serine residues in the proteolytic enzyme chymotrypsin, implying that this serine residue is especially reactive. As we will see in Chapter 9, it is indeed the case that this serine residue is at the active site. DIPF also revealed a reactive serine residue in acetylcholinesterase, an enzyme important in the transmission of nerve impulses. Thus, DIPF and similar compounds that bind and inactivate acetylcholinesterase are potent nerve gases.

Affinity labels are molecules that are structurally similar to the substrate for the enzyme that covalently modify active site residues. They are thus more specific for the enzyme active site than are group-specific reagents. Tosyl-l-phenylalanine chloromethyl ketone (TPCK) is a substrate analog for chymotrypsin. TPCK binds at the active site; and then reacts irreversibly with a histidine residue at that site, inhibiting the enzyme. The compound 3-bromoacetol is an affinity label for the enzyme triose phosphate isomerase (TIM). It mimics the normal substrate, dihydroxyacetone phosphate, by binding at the active site; then it covalently modifies the enzyme such that the enzyme is irreversibly inhibited.

Suicide inhibitors, or *mechanism-based inhibitors* are modified substrates that provide the most specific means to modify an enzyme active site. The inhibitor binds to the enzyme as a substrate and is initially processed by the normal catalytic mechanism. The mechanism of catalysis then generates a chemically reactive intermediate that inactivates the enzyme through covalent modification. The fact that the enzyme participates in its own irreversible inhibition strongly suggests that the covalently modified group on the enzyme is catalytically vital. One example of such an inhibitor is *N,N*-dimethylpropargylamine. A flavin prosthetic group of monoamine oxidase (MAO) oxidizes the *N,N*-dimethylpropargylamine, which in turn inactivates the enzyme by covalently modifying the flavin prosthetic group by alkylating N-5. Monoamine oxidase deaminates neurotransmitters such as dopamine and serotonin, lowering their levels in the brain. Parkinson disease is associated with low levels of dopamine, and depression is associated with low levels of serotonin. The drug deprenyl, which is used to treat Parkinson disease and depression, is a suicide inhibitor of monoamine oxidase.

TRANSITION-STATE ANALOGS ARE POTENT INHIBITORS OF ENZYMES

We turn now to compounds that provide the most intimate views of the catalytic process itself. Linus Pauling proposed in 1948 that compounds resembling the transition state of a catalyzed reaction should be very effective inhibitors of enzymes. These mimics are called *transition-state analogs*. The inhibition of proline racemase is an instructive example. *The racemization of proline proceeds through a transition state in which the tetrahedral α- carbon atom has become trigonal by loss of a proton.* In the trigonal form, all three bonds are in the same plane; C_α also carries a net negative charge. This symmetric carbanion can be reprotonated on one side to give the l isomer or on the other side to give the d isomer.

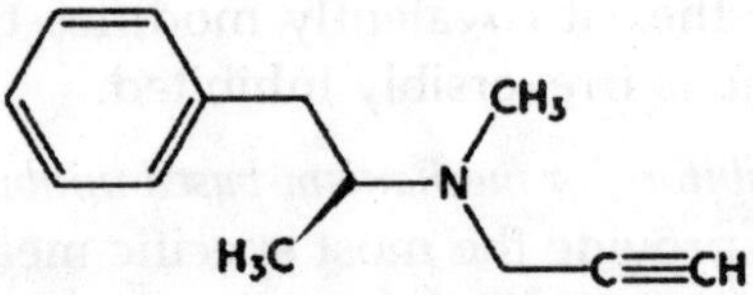

(–)Deprenyl

This picture is supported by the finding that the inhibitor pyrrole 2-carboxylate binds to the racemase 160 times as tightly as does proline. *The α-carbon atom of this inhibitor, like that of the transition state, is trigonal.* An analog that also carries a negative charge on C_α would be expected to bind even more tightly. In general, synthesizing compounds that more closely resemble the transition state than the substrate itself can produce highly potent and specific inhibitors of enzymes. The inhibitory power of transition-state analogs underscores the essence of catalysis: *selective binding of the transition state.*

CATALYTIC ANTIBODIES DEMONSTRATE THE IMPORTANCE OF SELECTIVE BINDING OF THE TRANSITION STATE TO ENZYMATIC ACTIVITY

Antibodies that recognize transition states should function as catalysts, if our understanding of the importance of the transition state to catalysis is correct. The preparation of an antibody that catalyzes the insertion of a metal ion into a porphyrin nicely illustrates the validity of this approach. Ferrochelatase, the final enzyme in the biosynthetic pathway for the production of heme, catalyzes the insertion of Fe^{2+} into protoporphyrin IX. The nearly planar porphyrin must be bent for iron to enter. The recently determined crystal structure of the ferrochelatase bound to a substrate analog confirms that the enzyme does indeed bend one of the pyrole rings, distorting it 36 degrees to insert the iron.

The problem was to find a transition-state analog for this metallation reaction that could be used as an antigen (immunogen) to generate an antibody. The solution came from the results of studies showing that an alkylated porphyrin, *N*-methylprotoporphyrin, is a potent inhibitor of ferrochelatase. This compound resembles the transition state because N-*alkylation forces the porphyrin to be bent*. Moreover, it was known that *N*-alkylporphyrins chelate metal ions 10^4 times as fast as their unalkylated counterparts do. Bending increases the exposure of the pyrrole nitrogen lone pairs of electrons to solvent, which facilitates metal in binding.

An antibody catalyst was produced with the use of an *N*-alkylporphyrin as the immunogen. The resulting antibody presumably distorts a planar porphyrinto facilitate the entry of a metal.On average, an antibody molecule metallated 80 porphyrin molecules per hour, a rate only 10-fold less than that of ferrochelatase and 2500-fold faster than the uncatalyzed reaction. *Catalytic antibodies (abzymes) can indeed be produced by using transition-state analogs as antigens.* Antibodies catalyzing many other kinds of chemical reactions—exemplified by ester and amide hydrolysis, amide-bond formation, transesterification, photoinduced cleavage, photoinduced dimerization, decarboxylation, and oxidization—have been produced with the use of similar strategies. The results of studies with transition-state analogs provide strong evidence that enzymes can function complementary in structure to the transition state. The power of transition-state analogs is now evident:

1. They are sources of insight into catalytic mechanisms,
2. They can serve as potent and specific inhibitors of enzymes, and
3. They can be used as immunogens to generate a wide range of novel catalysts.

PENICILLIN IRREVERSIBLY INACTIVATES A KEY ENZYME IN BACTERIAL CELL-WALL SYNTHESIS

Penicillin, the first antibiotic discovered, consists of a thiazolidine ring fused to a *β-lactam* ring, to which a variable R group is attached by a peptide bond. In benzyl penicillin, for example, R is a benzyl group. This structure can undergo a variety of rearrangements, and, in particular, the β-lactam ring is very labile. Indeed, this instability is closely tied to the antibiotic action of penicillin, as will be evident shortly.

How does penicillin inhibit bacterial growth? In 1957, Joshua Lederberg showed that bacteria ordinarily susceptible to penicillin could be grown in its presence if a hypertonic medium were used. The organisms obtained in this way, called *protoplasts,* are devoid of a cell wall and consequently lyse when transferred to a normal medium. Hence, penicillin was inferred to interfere with the synthesis of the bacterial cell wall. The cell-wall macromolecule, called a *peptidoglycan,* consists of linear polysaccharide chains that are cross-linked by short peptides. The enormous bag-shaped peptidoglycan confers mechanical support and prevents bacteria from bursting in response to their high internal osmotic pressure.

In 1965, James Park and Jack Strominger independently deduced that penicillin blocks the last step in cell-wall synthesis—namely, the crosslinking of different peptidoglycan strands. In the formation of the cell wall of *Staphylococcus aureus,* the amino group at one end of a pentaglycine chain attacks the peptide bond between two d-alanine residues in another peptide unit. A peptide bond is formed between glycine and one of the d-alanine residues; the other d-alanine residue is released. This cross-linking reaction is catalyzed by *glycopeptide transpeptidase.* Bacterial cell walls are unique in containing d amino acids, which form cross-links by a mechanism different from that used to synthesize proteins.

Penicillin inhibits the cross-linking transpeptidase by the Trojan horse stratagem. The transpeptidase normally forms an *acyl intermediate* with the penultimate d-alanine residue of the d-Ala-d-Ala peptide. This covalent acyl-enzyme intermediate then reacts with the amino group of the terminal glycine in another peptide to form the cross-link. Penicillin is welcomed into the active site of the transpeptidase because it mimics the d-Ala-d-Ala moiety of the normal substrate. Bound penicillin then forms a covalent bond with a serine residue at the active site of the enzyme. *This penicilloyl-enzyme does not react further. Hence, the transpeptidase is irreversibly inhibited and cell-wall synthesis cannot take place.*

Why is penicillin such an effective inhibitor of the transpeptidase? The highly strained, four-membered β-lactam ring of penicillin makes it especially reactive. On binding to the transpeptidase, the serine residue at the active site attacks the carbonyl carbon atom of the lactam ring to form the penicilloyl-serine derivative. Because the peptidase participates in its own inactivation, penicillin acts as a suicide inhibitor.

VITAMINS ARE OFTEN PRECURSORS TO COENZYMES

Earlier we considered the fact that many enzymes re- quire cofactors to be catalytically active. One class of these cofactors, termed coenzymes, consists of small organic molecules, many of which are derived from *vitamins.* Vitamins themselves are organic molecules that are needed in small amounts in the diets of some higher animals. These molecules serve the same roles in nearly all forms of life, but higher animals lost the capacity to synthesize them in the course of evolution. For instance, whereas *E. coli* can thrive on glucose and organic salts, human beings require at least 12 vitamins in the diet. The biosynthetic pathways for vitamins can be complex; thus, it is biologically more efficient to ingest vitamins than to synthesize the enzymes required to construct them from simple molecules. This efficiency comes at the cost of dependence on other organisms for chemicals essential for life. Indeed, vitamin deficiency can generate diseases in all organisms requiring these molecules. Vitamins can be grouped according to whether they are soluble in water or in nonpolar solvents.

WATER-SOLUBLE VITAMINS FUNCTION AS COENZYMES

Ascorbate, the ionized form of ascorbic acid, serves as a reducing agent (an antioxidant), as will be discussed shortly. The vitamin B series comprises components of coenzymes. Note that, in all cases except vitamin C, the vitamin must be modified before it can serve its function.

Vitamin deficiencies are capable of causing a variety of pathological conditions. However, many of the same symptoms can result from conditions other than lack of a vitamin. For this reason and because vitamins are required in relatively small amounts, pathological conditions resulting from vitamin deficiencies are often difficult to diagnose.

The requirement for vitamin C proved relatively straightforward to demonstrate. This water-soluble vitamin is not used as a coenzyme but is still required for the continued activity of proyl hydroxylase. This enzyme synthesizes 4-hydroxyproline, an amino acid that is required in collagen, the major connective tissue in vertebrates, but is rarely found anywhere else. How is this unusual amino acid formed and what is its role? The results of radioactive-labeling studies showed that proline residues on the amino side of glycine residues in nascent collagen chains become hydroxylated. The oxygen atom that becomes attached to C-4 of proline comes from molecular oxygen, O_2. The other oxygen atom of O_2 is taken up by α-ketoglutarate, which is converted into succinate.

This complex reaction is catalyzed by *prolyl hydroxylase*, a *dioxygenase*. It is assisted by an Fe^{2+} ion, which is tightly bound to it and needed to activate O_2. The enzyme also converts α-ketoglutarate into succinate without hydroxylating proline. In this partial reaction, an oxidized iron complex is formed, which inactivates the enzyme. How is the active enzyme regenerated? *Ascorbate (vitamin C)* comes to the rescue by reducing the ferric ion of the inactivated enzyme. In the recovery process, ascorbate is oxidized to dehydroascorbic acid. Thus, ascorbate serves here as a specific *antioxidant*.

Primates are unable to synthesize ascorbic acid and hence must acquire it from their diets. The importance of ascorbate becomes strikingly evident in *scurvy*. Jacques Cartier in 1536 gave a vivid description of this dietary deficiency disease, which afflicted his men as they were exploring the Saint Lawrence River:

Some did lose all their strength, and could not stand on their feet. ... Others also had all their skins spotted with spots of blood of a purple colour: then did it ascend up to their ankles, knees, thighs, shoulders, arms, and necks. Their mouths became stinking, their gums so rotten, that all the flesh did fall off, even to the roots of the teeth, which did also almost all fall out.

James Lind, a Scottish physician, illuminated the means of preventing scurvy in an article titled "A Treatise of the Scurvy" published in 1747. Lind described a controlled study establishing that scurvy could be prevented by including citrus fruits in the diet. The Royal Navy eventually began issuing lime rations to sailors, from which custom British sailors acquired the nickname "limeys." Lind's research was inspired by the plight of an expedition commanded by Commodore George Anson. Anson left England in 1740 with a fleet of six ships and more than 1000 men and returned with an enormous amount of treasure, but of his crew only 145 survived to reach home. The remainder had died of scurvy.

Why does impaired hydroxylation have such devastating consequences? *Collagen synthesized in the absence of ascorbate is less stable than the normal protein*. Studies of the thermal stability of synthetic polypeptides have been especially informative. Hydroxyproline stabilizes the collagen triple helix by forming interstrand hydrogen bonds. The abnormal fibers formed by insufficiently hydroxylated collagen contribute to the skin lesions and blood-vessel fragility seen in scurvy.

FAT-SOLUBLE VITAMINS PARTICIPATE IN DIVERSE PROCESSES SUCH AS BLOOD CLOTTING AND VISION

Not all vitamins function as coenzymes. The *fat-soluble vitamins*, which are designated by the letters A, D, E, and K have a diverse array of functions. Vitamin K, which is required for normal blood clotting (*K* from the German *k*oagulation), participates in the carboxylation of glutamate residues to γ-carboxyglutamate, which makes modified glutamic acid a much stronger chelator of Ca^{2+}. Vitamin A (retinol) is the precursor of retinal, the light-sensitive group in rhodopsin and other visual pigments. A deficiency of this vitamin leads to night blindness. In addition, young animals require vitamin A for growth. Retinoic acid, which contains a terminal carboxylate in place of the alcohol terminus of retinol, serves as a signal molecule and activates the transcription of specific genes that mediate growth and development. A metabolite of vitamin D is a hormone that regulates the metabolism of calcium and phosphorus. A deficiency in vitamin D impairs bone formation in growing animals. Infertility in rats is a consequence of vitamin E (α-tocopherol) deficiency. This vitamin reacts with and neutralizes reactive oxygen species such as hydroxyl, radicals before they can oxidize unsaturated membrane lipids, damaging cell structures.

Chapter 9

Catalytic Strategies

What are the sources of the catalytic power and specificity of enzymes? This chapter presents the catalytic strategies used by four classes of enzymes: the serine proteases, carbonic anhydrases, restriction endonucleases, and nucleoside monophosphate (NMP) kinases. The first three classes of enzymes catalyze reactions that require the addition of water to a substrate. For the serine proteases, exemplified by chymotrypsin, the challenge is to promote a reaction that is almost immeasurably slow at neutral pH in the absence of a catalyst. For carbonic anhydrases, the challenge is to achieve a high absolute rate of reaction, suitable for integration with other rapid physiological processes. For restriction endonucleases such as *Eco*RV, the challenge is to attain a very high level of specificity. Finally, for NMP kinases, the challenge is to transfer a phosphoryl group from ATP to a nucleotide and not to water. The actions of these enzymes illustrate many important principles of catalysis. The mechanisms of these enzymes have been revealed through the use of incisive experimental probes, including the techniques of protein structure determination and site-directed mutagenesis. These mechanisms include the use of binding energy and induced fit as well as several specific catalytic strategies. Properties common to an enzyme family reveal how their enzyme active sites have evolved and been refined. Structural and mechanistic comparisons of enzyme action are thus sources of insight into the evolutionary history of enzymes. These comparisons also reveal particularly effective solutions to biochemical problems that are used repeatedly in biological systems. In addition, our knowledge of catalytic strategies has been used to develop practical applications, including drugs that are potent and specific enzyme inhibitors. Finally, although we shall not consider catalytic RNA molecules explicitly in this chapter, the principles apply to these catalysts in addition to protein catalysts.

A FEW BASIC CATALYTIC PRINCIPLES ARE USED BY MANY ENZYMES

We learned that enzymatic catalysis begins with substrate binding. The *binding energy* is the free energy released in the formation of a large number of weak interactions between the enzyme and the substrate. We can envision this binding energy as serving two purposes: it establishes substrate specificity and increases catalytic efficiency. Only the correct substrate can participate in most or all of the interactions with the enzyme and thus maximize binding energy, accounting for the exquisite substrate specificity exhibited by many enzymes. Furthermore, the full complement of such interactions is formed only when the substrate is in the transition state. Thus, interactions between the enzyme and the substrate not only favour substrate binding but stabilize the transition state, thereby lowering the activation energy. The binding energy can also promote structural changes in both the enzyme and the substrate that facilitate catalysis, a process referred to as *induced fit*.

Enzymes commonly employ one or more of the following strategies to catalyze specific reactions:

1. *Covalent catalysis*. In covalent catalysis, the active site contains a reactive group, usually a powerful nucleophile that becomes temporarily covalently modified in the course of catalysis. The proteolytic enzyme chymotrypsin provides an excellent example of this mechanism.
2. *General acid-base catalysis*. In general acid-base catalysis, a molecule other than water plays the role of a proton donor or acceptor. Chymotrypsin uses a

histidine residue as a base catalyst to enhance the nucleophilic power of serine.

3. *Metal ion catalysis.* Metal ions can function catalytically in several ways. For instance, a metal ion may serve as an electrophilic catalyst, stabilizing a negative charge on a reaction intermediate. Alternatively, the metal ion may generate a nucleophile by increasing the acidity of a nearby molecule, such as water in the hydration of CO_2 by carbonic anhydrase. Finally, the metal ion may bind to substrate, increasing the number of interactions with the enzyme and thus the binding energy. This strategy is used by NMP kinases.
4. *Catalysis by approximation.* Many reactions include two distinct substrates. In such cases, the reaction rate may be considerably enhanced by bringing the two substrates together along a single binding surface on an enzyme. NMP kinases bring two nucleotides together to facilitate the transfer of a phosphoryl group from one nucleotide to the other.

PROTEASES: FACILITATING A DIFFICULT REACTION

Protein turnover is an important process in living systems. Proteins that have served their purpose must be degraded so that their constituent amino acids can be recycled for the synthesis of new proteins. Proteins ingested in the diet must be broken down into small peptides and amino acids for absorption in the gut. Furthermore, as described in detail in Chapter, proteolytic reactions are important in regulating the activity of certain enzymes and other proteins.

Proteases cleave proteins by a hydrolysis reaction—the addition of a molecule of water to a peptide bond:

$$R_1{-}C(=O){-}N(H){-}R_2 + H_2O \rightleftharpoons R_1{-}C(=O){-}O^- + R_2{-}NH_3^+$$

Although the hydrolysis of peptide bonds is thermodynamically favoured, such hydrolysis reactions are extremely slow. In the absence of a catalyst, the half-life for the hydrolysis of a typical peptide at neutral pH is estimated to be between 10 and 1000 years. Yet, peptide bonds must be hydrolyzed within milliseconds in some biochemical processes.

The chemical bonding in peptide bonds is responsible for their kinetic stability. Specifically, the resonance structure that accounts for the planarity of a peptide bond also makes such bonds resistant to hydrolysis. This resonance structure endows the peptide bond with partial double-bond character:

$$R_1{-}C({=}O){-}N(H){-}R_2 \longleftrightarrow R_1{-}C(O^-){=}N^+(H){-}R_2$$

The carbon-nitrogen bond is strengthened by its double-bond character, and the carbonyl carbon atom is less electrophilic and less susceptible to nucleophilic attack than are the carbonyl carbon atoms in compounds such as carboxylate esters. Consequently, to promote peptide-bond cleavage, an enzyme must facilitate nucleophilic attack at a normally unreactive carbonyl group.

CHYMOTRYPSIN POSSESSES A HIGHLY REACTIVE SERINE RESIDUE

A number of proteolytic enzymes participate in the breakdown of proteins in the digestive systems of mammals and other organisms. One such enzyme, chymotrypsin, cleaves peptide bonds selectively on the carboxylterminal side of the large hydrophobic amino acids such as tryptophan, tyrosine, phenylalanine, and methionine. Chymotrypsin is a good example of the use of *covalent modification* as a catalytic strategy. The enzyme employs a powerful nucleophile to attack the unreactive carbonyl group of the substrate. This nucleophile becomes covalently attached to the substrate briefly in the course of catalysis.

What is the nucleophile that chymotrypsin employs to attack the substrate carbonyl group? A clue came from the fact that chymotrypsin contains an extraordinarily reactive serine residue. Treatment with organofluorophosphates such as diisopropylphosphofluoridate (DIPF) was found to inactivate the enzyme irreversibly. Despite the fact that the enzyme possesses 28 serine residues, only one, serine 195, was modified, resulting in a total loss of enzyme activity. This *chemical modification reaction* suggested that this unusually reactive serine residue plays a central role in the catalytic mechanism of chymotrypsin.

CHYMOTRYPSIN ACTION PROCEEDS IN TWO STEPS LINKED BY A COVALENTLY BOUND INTERMEDIATE

How can we elucidate the role of serine 195 in chymotrypsin action? A study of the enzyme's kinetics provided a second clue to chymotrypsin's catalytic mechanism and the role of serine 195. The kinetics of enzyme action are often easily monitored by having the enzyme act on a substrate analog that forms a coloured product. For chymotrypsin, such a *chromogenic substrate* is *N*-acetyl-l-phenylalanine *p*-nitrophenyl ester. This substrate is an ester rather than an amide, but many proteases will also hydrolyze esters. One of the products formed by chymotrypsin's cleavage of this substrate is *p*- nitrophenolate, which has a yellow colour. Measurements of the absorbance of light revealed the amount of *p*-nitrophenolate being produced.

Under steady-state conditions, the cleavage of this substrate obeys Michaelis-Menten kinetics with a K_M of 20 μM and a k_{cat} of 77 s^{-1}. The initial phase of the reaction was examined by using the stopped-flow method. This technique permits the rapid mixing of enzyme and substrate and allows almost instantaneous monitoring of the reaction. At the beginning of the reaction, this method revealed a "burst" phase during which the coloured product was produced rapidly. Product was then produced more slowly as the reaction reached the steady state. These results suggest that hydrolysis proceeds in two steps. The burst is observed because, for this substrate, the first step is more rapid than the second step.

The two steps are explained by the reaction of the serine nucleophile with the substrate to form the covalently bound enzyme-substrate intermediate. First, the highly reactive serine 195 hydroxyl group attacks the carbonyl group of the substrate to form the acyl-enzyme intermediate, releasing the alcohol *p*-nitrophenol (or an amine if the substrate is an amide rather than an ester). Second, the acyl-enzyme intermediate is hydrolyzed to release the carboxylic acid component of the substrate and regenerate the free enzyme. Thus, *p*-nitrophenolate is produced rapidly on the addition of the substrate as the acyl-enzyme intermediate is formed, but it takes longer for the enzyme to be "reset" by the hydrolysis of the acyl-enzyme intermediate.

SERINE IS PART OF A CATALYTIC TRIAD THAT ALSO INCLUDES HISTIDINE AND ASPARTIC ACID

The determination of the three-dimensional structure of chymotrypsin by David Blow in 1967 was a source of further insight into its mechanism of action. Overall, chymotrypsin is roughly spherical and comprises three polypeptide chains, linked by disulfide bonds. It is synthesized as a single polypeptide, termed chymotrypsinogen, which is activated by the proteolytic cleavage of the polypeptide to yield the three chains. The active site of chymotrypsin, marked by serine 195, lies in a cleft on the surface of the enzyme. The structural analysis revealed the chemical basis of the special reactivity of serine 195. The side chain of serine 195 is hydrogen bonded to the imidazole ring of histidine 57. The -NH group of this imidazole ring is, in turn, hydrogen bonded to the carboxylate group of aspartate 102. This constellation of residues is referred to as the *catalytic triad*. How does this arrangement of residues lead to the high reactivity of serine 195? The histidine residue serves to position the serine side chain and to polarize its hydroxyl group. In doing so, the residue acts as a general base catalyst, a hydrogen ion acceptor, because the polarized hydroxyl group of the serine residue is poised for deprotonation. The withdrawal of the proton from the hydroxyl group generates an alkoxide ion, which is a much more powerful nucleophile than an alcohol is. The aspartate residue helps orient the histidine residue and make it a better proton acceptor through electrostatic effects.

These observations suggest a mechanism for peptide hydrolysis. After substrate binding (step 1), the reaction begins with the hydroxyl group of serine 195 making a nucleophilic attack on the carbonyl carbon atom of the substrate (step 2). The nucleophilic attack changes the geometry around this carbon atom from trigonal planar to tetrahedral. The inherently unstable *tetrahedral intermediate* formed bears a formal negative charge on the oxygen atom derived from the carbonyl group. This charge is stabilized by interactions with NH groups from the protein in a site termed the *oxyanion hole*. These interactions also help stabilize the transition state that precedes the formation of the tetrahedral intermediate. This tetrahedral intermediate then collapses to generate the acyl-enzyme (step 3). This step is facilitated by the transfer of a proton from the positively charged histidine residue to the amino group formed by cleavage of the peptide bond. The amine component is now free to depart from the enzyme (step 4) and is replaced by a water molecule (step 5). The ester group of the acyl-enzyme is now hydrolyzed by a process that is essentially a repeat of steps 2 through 4. The water molecule attacks the carbonyl group while a proton is concomitantly removed by the histidine residue, which now acts as a general acid catalyst, forming a tetrahedral intermediate (step 6). This structure breaks down to form the carboxylic acid product (step 7). Finally, the release of the carboxylic acid product (step 8) readies the enzyme for another round of catalysis.

This mechanism accounts for all characteristics of chymotrypsin action except the observed preference for cleaving the peptide bonds just past residues with large, hydrophobic side chains. Examination of the threedimensional structure of chymotrypsin with substrate analogs and enzyme inhibitors revealed the presence of a deep, relatively hydrophobic pocket, called the S_1 pocket, into which the long, uncharged side chains of residues such as phenylalanine and tryptophan can fit. *The binding of an appropriate side chain into this pocket positions the adjacent peptide bond into the active site for cleavage.* The specificity of chymotrypsin depends almost entirely on which amino acid is directly on the amino-terminal side of the peptide bond to be cleaved. Other proteases have more-complex specificity patterns. Such enzymes have additional pockets on their surfaces for the recognition of other residues in the substrate. Residues on the amino-terminal side of the scissile bond (the bond to be cleaved) are labeled P_1, P_2, P_3, and so forth, indicating their positions in relation to the scissile bond. Likewise, residues on the carboxyl side of the scissile bond are labeled P_1', P_2', P_3', and so forth. The corresponding sites on the enzyme are referred to as S_1, S_2 or S_1', S_2', and so forth.

CATALYTIC TRIADS ARE FOUND IN OTHER HYDROLYTIC ENZYMES

Many other proteins have subsequently been found to contain catalytic triads similar to that discovered in chymotrypsin. Some, such as trypsin and elastase, are obvious homologs of chymotrypsin. The sequences of these proteins are approximately 40per cent identical with that of chymotrypsin, and their overall structures are nearly the same. These proteins operate by mechanisms identical with that of chymotrypsin. However, they have very different substrate specificities. Trypsin cleaves at the peptide bond after residues with long, positively charged side chains—namely, arginine and lysine—whereas elastase cleaves at the peptide bond after amino acids with small side chains—such as alanine and serine. Comparison of the S_1 pockets of these enzymes reveals the basis of the specificity. In trypsin, an aspartate residue (Asp 189) is present at the bottom of the S_1 pocket in place of a serine residue in chymotrypsin. The aspartate residue attracts and stabilizes a positively charged arginine or lysine residue in the substrate. In elastase, two residues at the top of the pocket in chymotrypsin and trypsin are replaced with valine (Val 190 and Val 216). These residues close off the mouth of the pocket so that only small side chains may enter.

Other members of the chymotrypsin family include a collection of proteins that take part in blood clotting. In addition, a wide range of proteases found in bacteria and viruses also belong to this clan. Furthermore, other enzymes that are not homologs of chymotrypsin have been found to contain very similar active sites. As noted the presence of very similar active sites in these different protein families is a consequence of convergent evolution. Subtilisin, a protease in bacteria such as *Bacillus amyloliquefaciens,* is a particularly well characterized example. The active site of this enzyme includes both the catalytic triad and the oxyanion hole. However, one of the NH groups that forms the oxyanion hole comes from the side chain of an asparagine residue rather than from the peptide backbone. Subtilisin is the founding member of another large family of proteases that includes representatives from Archaea, Eubacteria, and Eukarya.

Yet another example of the catalytic triad has been found in carboxypeptidase II from wheat. The structure of this enzyme is not significantly similar to either chymotrypsin or subtilisin. This protein is a member of an intriguing family of homologous proteins that includes esterases such as acetylcholine esterase and certain lipases. These enzymes all make use of histidine-activated nucleophiles, but the nucleophiles may be cysteine rather than serine. Finally, other proteases have been discovered that contain an active-site serine or threonine residue that is activated not by a histidine-aspartate pair but by a primary amino group from the side chain of lysine or by the N-terminal amino group of the polypeptide chain.

Thus, the catalytic triad in proteases has emerged at least three times in the course of evolution. We can conclude that this catalytic strategy must be an especially effective approach to the hydrolysis of peptides and related bonds.

THE CATALYTIC TRIAD HAS BEEN DISSECTED BY SITE-DIRECTED MUTAGENESIS

In particular, site-directed mutagenesis has been used to test the contribution of individual amino acid residues to the catalytic power of an enzyme. Subtilisin has been extensively studied by this method. Each of the residues within the catalytic triad, consisting of aspartic acid 32, histidine 64, and serine 221, has been individually converted into alanine, and the ability of each mutant enzyme to cleave a model substrate has been examined. As expected, the conversion of active-site serine 221 into alanine dramatically reduced catalytic power; the value of k_{cat} fell to less than *one-millionth* of its value for the wild-type enzyme. The value of K_M was essentially unchanged: its increase by no more than a factor of two indicated that substrate binding is not significantly affected. The mutation of histidine 64 to alanine had very similar effects. These observations support the notion that the serine-histidine pair act together to generate a nucleophile of sufficient power to attack the carbonyl group of a peptide bond. The conversion of aspartate 32 into alanine had a smaller effect, although the value of k_{cat} still fell to less than 0.005per cent of its wild-type value. The simultaneously conversion of all three catalytic triad residues into alanine was no more deleterious than the conversion of serine or histidine alone. Despite the reduction in their catalytic power, the mutated enzymes still hydrolyze peptides a thousand times as rapidly as does buffer at pH 8.6.

Because the oxyanion hole of subtilisin includes a side-chain NH group in addition to backbone NH groups, it is possible to probe the importance of the oxyanion hole for catalysis by site-directed mutagenesis. The mutation of asparagine 155 to glycine reduced the value of k_{cat} to 0.2per cent of its wild-type value but increased the value of K_M by only a factor of two. These observations demonstrate that the NH group of the asparagine residue plays a significant role in stabilizing the tetrahedral intermediate and the transition state leading to it.

CYSTEINE, ASPARTYL, AND METALLOPROTEASES ARE OTHER MAJOR CLASSES OF PEPTIDE-CLEAVING ENZYMES

Not all proteases utilize strategies based on activated serine residues.

Classes of proteins have been discovered that employ three alternative approaches to peptide-bond hydrolysis. These classes are the

1. Cysteine proteases,
2. Aspartyl proteases, and
3. Metalloproteases. In each case, the strategy generates a nucleophile that attacks the peptide carbonyl group.

The strategy used by the *cysteine proteases* is most similar to that used by the chymotrypsin family. In these enzymes, a cysteine residue, activated by a histidine residue, plays the role of the nucleophile that attacks the peptide bond in a manner quite analogous to that of the serine residue in serine proteases. An ideal example of these proteins is papain, an enzyme purified from the fruit of the papaya. Mammalian proteases homologous to papain have been discovered, most notably the cathepsins, proteins having a role in the immune and other systems. The cysteine-based active site arose independently at least twice in the course of evolution; the caspases, enzymes that play a major role in apoptosis have active sites similar to that of papain, but their overall structures are unrelated.

The second class comprises the *aspartyl proteases*. The central feature of the active sites is a pair of aspartic acid residues that act together to allow a water molecule to attack the peptide bond. One aspartic acid residue (in its deprotonated form) activates the attacking water molecule by poising it for deprotonation, whereas the other aspartic acid residue (in its protonated form) polarizes the peptide carbonyl, increasing its susceptibility to attack. Members of this class include renin, an enzyme having a role in the regulation of blood pressure, and the digestive enzyme pepsin. These proteins possess approximate twofold symmetry, suggesting that the two halves are evolutionarily related. A likely scenario is that two copies of a gene for the ancestral enzyme fused to form a single gene that encoded a single-chain enzyme. Each copy of the gene would have contributed an aspartate residue to the active site. The human immunodeficiency virus (HIV) and other retroviruses contain an unfused dimeric aspartyl protease that is similar to the fused protein, but the individual chains are not joined to make a single chain. This observation is consistent with the idea that the enzyme may have originally existed as separate subunits.

The *metalloproteases* constitute the final major class of peptide-cleaving enzymes. The active site of such a protein contains a bound metal ion, almost always zinc, that activates a water molecule to act as a nucleophile to attack the peptide carbonyl group. The bacterial enzyme thermolysin and the digestive enzyme carboxypeptidase A are classic examples of the zinc proteases. Thermolysin, but not carboxypeptidase A, is a member of a large and diverse family of homologous zinc proteases that includes the matrix metalloproteases, enzymes that catalyze the reactions in tissue remodeling and degradation.

In each of these three classes of enzymes, the active site includes features that allow for the activation of water or another nucleophile as well as for the polarization of the peptide carbonyl group and subsequent stabilization of a tetrahedral intermediate.

PROTEASE INHIBITORS ARE IMPORTANT DRUGS

Compounds that block or modulate the activities of proteases can have dramatic biological effects. Most natural *protease inhibitors* are similar in structure to the peptide substrates of the enzyme that each inhibits. *Several important drugs are protease inhibitors.* For example, captopril, an inhibitor of the metalloprotease angiotensin-converting enzyme (ACE), has been used to regulate blood pressure. Crixivan, an inhibitor of the HIV protease, is used in the treatment of AIDS. This protease cleaves multidomain viral proteins into their active forms; blocking this process completely prevents the virus from being infectious. To prevent unwanted side effects, protease inhibitors used as drugs must be specific for one enzyme without inhibiting other proteins within the body.

Let us examine the interaction of Crixivan with HIV protease in more detail. Crixivan is constructed around an alcohol that mimics the tetrahedral intermediate; other groups are present to bind into the S_2, S_1, S_1', and S_2' recognition sites on the enzyme. The results of x-ray crystallographic studies revealed the structure of the enzyme-Crixivan complex, showing that Crixivan adopts a conformation that approximates the twofold symmetry of the enzyme. The active site of HIV protease is covered by two apparently flexible flaps that fold down on top of the bound inhibitor. The hydroxyl group of the central alcohol interacts with two aspartate residues of the active site, one in each subunit. In addition, two carbonyl groups of the inhibitor are hydrogen bonded to a water molecule (not shown), which, in turn, is hydrogen bonded to a peptide NH group in each of the flaps. This interaction of the inhibitor with water and the enzyme is not possible with cellular aspartyl proteases such as renin and thus may contribute to the specificity of Crixivan and other inhibitors for HIV protease.

MAKING A FAST REACTION FASTER: CARBONIC ANHYDRASES

Carbon dioxide is a major end product of aerobic metabolism. In complex organisms, this carbon dioxide is released into the blood and transported to the lungs for exhalation. While in the blood, carbon dioxide reacts with water. The product of this reaction is a moderately strong acid, carbonic acid (pK_a = 3.5), which becomes bicarbonate ion on the loss of a proton.

$$O{=}C{=}O + H_2O \underset{k_{-1}}{\overset{k_1}{\rightleftharpoons}} HO{-}C({=}O){-}OH \rightleftharpoons HO{-}C({=}O){-}O^- + H^+$$

Carbonic acid   Bicarbonate ion

Even in the absence of a catalyst, this hydration reaction proceeds at a moderate pace. At 37°C near neutral pH, the second-order rate constant k_1 is 0.0027 M^{-1} s^{-1}. This corresponds to an effective first-order rate constant of 0.15 s^{-1}

in water ($[H_2O] = 55.5$ M). Similarly, the reverse reaction, the dehydration of bicarbonate, is relatively rapid, with a rate constant of $k_{-1} = 50\ s^{-1}$. These rate constants correspond to an equilibrium constant of $K_1 = 5.4 \times 10^{-5}$ and a ratio of $[CO_2]$ to $[H_2CO_3]$ of 340:1.

Despite the fact that CO_2 hydration and HCO_3^- dehydration occur spontaneously at reasonable rates in the absence of catalysts, almost all organisms contain enzymes, referred to as *carbonic anhydrases,* that catalyze these processes. Such enzymes are required because CO_2 hydration and HCO_3^- dehydration are often coupled to rapid processes, particularly transport processes. For example, HCO_3^- in the blood must be dehydrated to form CO_2 for exhalation as the blood passes through the lungs. Conversely, CO_2 must be converted into HCO_3^- for the generation of the aqueous humour of the eye and other secretions. Furthermore, both CO_2 and HCO_3^- are substrates and products for a variety of enzymes, and the rapid interconversion of these species may be necessary to ensure appropriate substrate levels. So important are these enzymes in human beings that mutations in some carbonic anhydrases have been found to cause osteopetrosis (excessive formation of dense bones accompanied by anemia) and mental retardation.

Carbonic anhydrases accelerate CO_2 hydration dramatically. The most active enzymes, typified by human carbonic anhydrase II, hydrate CO_2 at rates as high as $k_{cat} = 10^6\ s^{-1}$, or a million times a second. Fundamental physical processes such as diffusion and proton transfer ordinarily limit the rate of hydration, and so special strategies are required to attain such prodigious rates.

CARBONIC ANHYDRASE CONTAINS A BOUND ZINC ION ESSENTIAL FOR CATALYTIC ACTIVITY

Less than 10 years after the discovery of carbonic anhydrase in 1932, this enzyme was found to contain bound zinc, associated with catalytic activity. This discovery, remarkable at the time, made carbonic anhydrase the first known zinc-containing enzyme. At present, hundreds of enzymes are known to contain zinc. In fact, more than one-third of all enzymes either contain bound metal ions or require the addition of such ions for activity. The chemical reactivity of metal ions—associated with their positive charges, with their ability to form relatively strong yet kinetically labile bonds, and, in some cases, with their capacity to be stable in more than one oxidation state—explains why catalytic strategies that employ metal ions have been adopted throughout evolution.

The results of x-ray crystallographic studies have supplied the most detailed and direct information about the zinc site in carbonic anhydrase. At least seven carbonic anhydrases, each with its own gene, are present in human beings. They are all clearly homologous, as revealed by substantial levels of sequence identity. Carbonic anhydrase II, present in relatively high concentrations in red blood cells, has been the most extensively studied.

Zinc is found only in the + 2 state in biological systems; so we need consider only this oxidation level as we examine the mechanism of carbonic anhydrase. A zinc atom is essentially always bound to four or more ligands; in carbonic anhydrase, three coordination sites are occupied by the imidazole rings of three histidine residues and an additional coordination site is occupied by a water molecule (or hydroxide ion, depending on pH). Because all of the molecules occupying the coordination sites are neutral, the overall charge on the $Zn(His)_3$ unit remains +2.

CATALYSIS ENTAILS ZINC ACTIVATION OF WATER

How does this zinc complex facilitate carbon dioxide hydration? A major clue comes from the pH profile of enzymatically catalyzed carbon dioxide hydration. At pH 8, the reaction proceeds near its maximal rate. As the pH decreases, the rate of the reaction drops. The midpoint of this transition is near pH 7, suggesting that a group with $pK_a = 7$ plays an important role in the activity of carbonic anhydrase and that the deprotonated (high pH) form of this group participates more effectively in catalysis. Although some amino acids, notably histidine, have pK_a values near 7, *a variety of evidence suggests that the group responsible for this transition is not an amino acid but is the zinc-bound water molecule.* Thus, the binding of a water molecule to the positively charged zinc centre reduces the pK_a of the water molecule from 15.7 to 7. With the lowered pK_a, a substantial concentration of hydroxide ion (bound to zinc) is generated at neutral pH. A zinc-bound hydroxide ion is sufficiently nucleophilic to attack carbon dioxide much more readily than water does. The importance of the zinc-bound hydroxide ion suggests a simple mechanism for carbon dioxide hydration.

1. Zinc facilitates the release of a proton from a water molecule, which generates a hydroxide ion.
2. The carbon dioxide substrate binds to the enzyme's active site and is positioned to react with the hydroxide ion.
3. The hydroxide ion attacks the carbon dioxide, converting it into bicarbonate ion.
4. The catalytic site is regenerated with the release of the bicarbonate ion and the binding of another molecule of water.

Thus, the binding of water to zinc favours the formation of the transition state, leading to bicarbonate formation by facilitating proton release and by bringing the two reactants into close proximity. A range of studies supports this mechanism. In particular, studies of a *synthetic analog model system* provide evidence for its plausibility. A simple synthetic ligand binds zinc through four nitrogen atoms (compared with three histidine nitrogen atoms in the enzyme. One water molecule remains bound to the zinc ion in the complex. Direct measurements reveal that this water molecule has a pK_a value of 8.7, not as low as the value for the water molecule in carbonic anhydrase but substantially lower than the value for free water. At pH 9.2, this complex accelerates the hydration of carbon dioxide more than 100-fold. Although catalysis by this synthetic system is much less efficient than catalysis by carbonic anhydrase, the model system strongly suggests that the zinc-bound hydroxide mechanism is likely to be correct. Carbonic anhydrases have evolved to utilize the reactivity intrinsic to a zinc-bound hydroxide ion as a potent catalyst.

A PROTON SHUTTLE FACILITATES RAPID REGENERATION OF THE ACTIVE FORM OF THE ENZYME

As noted earlier, some carbonic anhydrases can hydrate carbon dioxide at rates as high as a million times a second (10^6 s^{-1}). The magnitude of this rate can be understood from the following observations. At the conclusion of a carbon dioxide hydration reaction, the zinc-bound water molecule must lose a proton to regenerate the active form of the enzyme. The rate of the reverse reaction, the protonation of the zinc-bound hydroxide ion, is limited by the rate of proton diffusion. Protons diffuse very rapidly with second-order rate constants near 10^{-11} M^{-1} s^{-1}. Thus, the backward rate constant k_{-1} must be less than 10^{11} M^{-1} s^{-1}. Because the equilibrium constant K is equal to k_1/k_{-1}, the forward rate constant is given by $k_1 = K \cdot k_{-1}$. Thus, if $k_{-1} \leq 10^{11}$ M^{-1} s^{-1} and $K = 10^{-7}$ M (because $pK_a = 7$), then k_1 must be less than or equal to 10^4 s^{-1}. In other words, the rate of proton diffusion limits the rate of proton release to less than 10^4 s^{-1} for a group with $pK_a = 7$. However, if carbon dioxide is hydrated at a rate of 10^6 s^{-1}, then every step in the mechanism must take place at least this fast. How can this apparent paradox be resolved?

The answer became clear with the realization that *the highest rates of carbon dioxide hydration require the presence of buffer, suggesting that the buffer components participate in the reaction.* The buffer can bind or release protons. The advantage is that, whereas the concentrations of protons and hydroxide ions are limited to 10^{-7} M at neutral pH, the concentration of buffer components can be much higher, on the order of several millimolar.If the buffer component BH^+ has a pK_a of 7 (matching that for the zinc-bound water), then the equilibrium constant for the reaction is 1. The rate of proton abstraction is given by $k_1' \cdot [B]$. The second-order rate constants k_1' and k_{-1}' will be limited by buffer diffusion to values less than approximately 10^9 M^{-1} s^{-1}. Thus, buffer concentrations greater than [B] = 10^{-3} M (1 mM) may be high enough to support carbon dioxide hydration rates of 10^6 M^{-1} s^{-1} because $k_1' \cdot [B] = (10^9\ M^{-1}\ s^{-1}) \cdot (10^{-3}\ M) = 10^6\ s^{-1}$. This prediction is confirmed experimentally.

The molecular components of many buffers are too large to reach the active site of carbonic anhydrase. Carbonic anhydrase II has evolved a *proton shuttle* to allow buffer components to participate in the reaction from solution. The primary component of this shuttle is histidine 64. This residue transfers protons from the zinc-bound water molecule to the protein surface and then to the buffer. Thus, catalytic function has been enhanced through the evolution of an apparatus for controlling proton transfer from and to the active site. Because protons participate in many biochemical reactions, the manipulation of the proton inventory within active sites is crucial to the function of many enzymes and explains the prominence of acid-base catalysis.

CONVERGENT EVOLUTION HAS GENERATED ZINC-BASED ACTIVE SITES IN DIFFERENT CARBONIC ANHYDRASES

Carbonic anhydrases homologous to the human enzymes, referred to as *α-carbonic anhydrases,* are common in animals and in some bacteria and algae. In addition, two other families of carbonic anhydrases have been discovered. The *β-carbonic anhydrases* are found in higher plants and in many bacterial species, including *E. coli.* These proteins contain the zinc required for catalytic activity but are not significantly similar in sequence to the α-carbonic anhydrases. Furthermore, the β-carbonic anhydrases have only one conserved histidine residue, whereas the α- carbonic anhydrases have three. No three-dimensional structure is yet available, but spectroscopic studies suggest that the zinc is bound by one histidine residue, two cysteine residues (conserved among β-carbonic anhydrases), and a water molecule. A third family, the *γ-carbonic anhydrases,* also has been identified, initially in the archaeon *Methanosarcina thermophila.* The crystal structure of this enzyme reveals three zinc sites extremely similar to those in the α-carbonic anhydrases. In this case, however, the three zinc sites lie at the interfaces between the three subunits of a trimeric enzyme. The very striking left-handed β-helix (a β strand twisted into a left-handed helix) structure present in this enzyme has also been found in enzymes that catalyze reactions unrelated to those of carbonic anhydrase. Thus, convergent evolution has generated carbonic anhydrases that rely on coordinated zinc ions at least three times. In each case, the catalytic activity appears to be associated with zinc-bound water molecules.

RESTRICTION ENZYMES: PERFORMING HIGHLY SPECIFIC DNA-CLEAVAGE REACTIONS

Let us next consider a hydrolytic reaction that results in the cleavage of DNA. Bacteria and archaea have evolved mechanisms to protect themselves from viral infections. Many viruses inject their DNA genomes into cells; once inside, the viral DNA hijacks the cell's machinery to drive the production of viral proteins and, eventually, of progeny virus. Often, a viral infection results in the death of the host. A major protective strategy for the host is to use *restriction endonucleases* (restriction enzymes) to degrade the viral DNA on its introduction into a cell. These enzymes recognize particular base sequences, called *recognition sequences* or *recognition sites,* in their target DNA and cleave that DNA at defined positions. The most well studied class of restriction enzymes comprises the so-called type II restriction enzymes, which cleave DNA *within* their recognition sequences. Other types of restriction enzymes cleave DNA at positions somewhat distant from their recognition sites.

Restriction endonucleases must show tremendous specificity at two levels. First, they must cleave only DNA molecules that contain recognition sites (hereafter referred to as *cognate DNA*) without cleaving DNA molecules that lack these sites. Suppose that a recognition sequence is six base pairs long. Because there are 4^6, or 4096, sequences having six base pairs, the concentration of sites that must not be cleaved will be approximately 5000-fold as high as the concentration of sites that should be cleaved. Thus, to keep from damaging host-cell DNA, endonucleases must cleave cognate DNA molecules much more than 5000 times as efficiently as they cleave nonspecific sites. Second, restriction enzymes must not degrade the host DNA. How do these enzymes manage to degrade viral DNA while sparing their own?

The restriction endonuclease *Eco*RV (from *E. coli*) cleaves double-stranded viral DNA molecules that contain the sequence 5′ -GATATC-3′ but leaves intact host DNA containing hundreds of such sequences. The host DNA is protected by other enzymes called methylases, which methylate adenine bases within host recognition sequences. For each restriction endonuclease, the host cell produces a corresponding methylase that marks the host DNA and prevents its degradation. These pairs of enzymes are referred to as *restriction-modification systems*. We shall return to the mechanism used to achieve the necessary levels of specificity after considering the chemistry of the cleavage process.

CLEAVAGE IS BY IN-LINE DISPLACEMENT OF 3′ OXYGEN FROM PHOSPHORUS BY MAGNESIUM-ACTIVATED WATER

The fundamental reaction catalyzed by restriction endonucleases is the hydrolysis of the phosphodiester backbone of DNA. Specifically, the bond between the 3′ oxygen atom and the phosphorus atom is broken. The products of this reaction are DNA strands with a free 3′ -hydroxyl group and a 5′ -phosphoryl group. This reaction proceeds by nucleophilic attack at the phosphorus atom. We will consider two types of mechanism, as suggested by analogy with the proteases. The restriction endonuclease might cleave DNA in mechanism 1 through a covalent intermediate, employing a potent nucleophile (Nu), or in mechanism 2 by direct hydrolysis:

Mechanism Type 1 (covalent intermediate)

$$R_2O\text{-}P(O)(O^-)\text{-}OR_1 + \text{enzyme—NuH} \rightleftharpoons \text{enzyme—Nu-}P(O)(O^-)\text{-}OR_2 + R_1OH$$

$$\text{enzyme—Nu-}P(O)(O^-)\text{-}OR_2 + H_2O \rightleftharpoons \text{enzyme—NuH} + R_2O\text{-}P(O)(O^-)\text{-}OH$$

Mechanism Type 2 (direct hydrolysis)

$$R_2O\text{-}P(O)(O^-)\text{-}OR_1 + H_2O \rightleftharpoons R_1OH + HO\text{-}P(O)(O^-)\text{-}OR_2$$

Each postulates a different nucleophile to carry out the attack on the phosphorus. In either case, each reaction takes place by an *in-line displacement* path:

$$Nu + (R_1O)(R_2O)(R_3O)P\text{—}L \rightleftharpoons \left[Nu\cdots P(OR_1)(R_2O)(OR_3)\cdots L\right] \rightleftharpoons N\text{—}P(OR_1)(OR_2)(OR_3) + L$$

The incoming nucleophile attacks the phosphorus atom, and a pentacoordinate transition state is formed. This species has a trigonal bipyramidal geometry centred at the phosphorus atom, with the incoming nucleophile at one apex of the two pyramids and the group that is displaced (the leaving group, L) at the other apex. The two mechanisms differ in the number of times the displacement occurs in the course of the reaction.

In the first type of mechanism, a nucleophile in the enzyme (analogous to serine 195 in chymotrypsin) attacks the phosphoryl group to form a covalent intermediate. In a second step, this intermediate is hydrolyzed to produce the final products. Because two displacement reactions take place at the phosphorus atom in the first mechanism, the stereochemical configuration at the phosphorus atom would be inverted and then inverted again, and the overall configuration would be retained. In the second type of mechanism, analogous to that used by the aspartyl and metalloproteases, an activated water molecule attacks the phosphorus atom directly. In this mechanism, a single displacement reaction takes place at the phosphorus atom. Hence, the stereochemical configuration of the tetrahedral phosphorus atom is inverted each time a displacement reaction takes place. Monitoring the stereochemical changes of the phosphorus could be one approach to determining the mechanism of restriction endonuclease action.

A difficulty is that the phosphorus centres in DNA are not chiral, because two of the groups bound to the phosphorus atom are simple oxygen atoms, identical with each other. This difficulty can be circumvented by preparing DNA molecules that contain chiral phosphoryl groups, made by replacing one oxygen atom with sulfur (called a phosphorothioate). Let us consider *Eco*RV endonuclease. This enzyme cleaves the phosphodiester bond between the T and the A at the centre of the recognition sequence 5′ -GATATC-3′. The first step in monitoring the activity of the enzyme is to synthesize an appropriate substrate for *Eco*RV containing phosphorothioates at the sites of cleavage. The reaction is then performed in water that has been greatly enriched in ^{18}O to allow the incoming oxygen atom to be marked. The location of the ^{18}O label with respect to the sulfur atom indicates whether the reaction proceeds with inversion or retention of stereochemistry. *The analysis revealed that the stereochemical configuration at the phosphorus atom was inverted only once with cleavage.* This result is consistent with a direct attack of water at phosphorus and rules out the formation of any covalently bound intermediate.

RESTRICTION ENZYMES REQUIRE MAGNESIUM FOR CATALYTIC ACTIVITY

Restriction endonucleases as well as many other enzymes that act on phosphate-containing substrates require Mg^{2+} or some other similar divalent cation for activity. What is the function of this metal?

It has been possible to examine the interactions of the magnesium ion when it is bound to the enzyme. Crystals have been produced of *Eco*RV endonuclease bound to oligonucleotides that contain the appropriate recognition sequences. These crystals are grown in the absence of magnesium to prevent cleavage; then, when produced, the crystals are soaked in solutions containing the metal. No cleavage takes place, allowing the location of the magnesium ion binding sites to be determined. The magnesium ion was found to be bound to six ligands: three are water molecules, two are carboxylates of the enzyme's aspartate residues, and one is an oxygen atom of the phosphoryl group at the site of cleavage. The magnesium ion holds a water molecule in a position from which the water molecule can attack the phosphoryl group and, in conjunction with the aspartate residues, helps polarize the water molecule toward deprotonation. Because cleavage does not take place within these crystals, the observed structure cannot be the true catalytic conformation. Additional studies have revealed that a second magnesium ion must be present in an adjacent site for *Eco*RV endonuclease to cleave its substrate.

THE COMPLETE CATALYTIC APPARATUS IS ASSEMBLED ONLY WITHIN COMPLEXES OF COGNATE DNA MOLECULES, ENSURING SPECIFICITY

We now return to the question of specificity, the defining feature of restriction enzymes. The recognition sequences for most restriction endonucleases are *inverted repeats.* This arrangement gives the three-dimensional structure of the recognition site a *twofold rotational symmetry*. The restriction enzymes display a corresponding symmetry to facilitate recognition: they are dimers whose two subunits are related by twofold rotational symmetry. The matching symmetry of the recognition sequence and the enzyme has been confirmed by the determination of the structure of the complex between *Eco*RV endonuclease and DNA fragments containing its recognition sequence. The enzyme surrounds the DNA in a tight embrace. Examination of this structure reveals features that are highly significant in determining specificity.

A unique set of interactions occurs between the enzyme and a cognate DNA sequence. Within the 5′ -GATATC-3′

sequence, the G and A bases at the 5′ end of each strand and their Watson-Crick partners directly contact the enzyme by hydrogen bonding with residues that are located in two loops, one projecting from the surface of each enzyme subunit. The most striking feature of this complex is the *distortion of the DNA*, which is substantially kinked in the centre. The central two TA base pairs in the recognition sequence play a key role in producing the kink. They do not make contact with the enzyme but appear to be required because of their ease of distortion. 5′ -TA-3′ sequences are known to be among the most easily deformed base pairs. The distortion of the DNA at this site has severe effects on the specificity of enzyme action.

Specificity is often determined by an enzyme's binding affinity for substrates. In regard to *Eco*RV endonuclease, however, binding studies performed in the absence of magnesium have demonstrated that the enzyme binds to all sequences, both cognate and noncognate, with approximately equal affinity. However, the structures of complexes formed with noncognate DNA fragments are strikingly different from those formed with cognate DNA: the noncognate DNA conformation is not substantially distorted. *This lack of distortion has important consequences with regard to catalysis. No phosphate is positioned sufficiently close to the active-site aspartate residues to complete a magnesium ion binding site.* Hence, the nonspecific complexes do not bind the magnesium ion and the complete catalytic apparatus is never assembled. The distortion of the substrate and the subsequent binding of the magnesium ion account for the catalytic specificity of more than 1,000,000-fold that is observed for *Eco*RV endonuclease despite very little preference at the level of substrate binding.

We can now see the role of binding energy in this strategy for attaining catalytic specificity. In binding to the enzyme, the DNA is distorted in such a way that additional contacts are made between the enzyme and the substrate, increasing the binding energy. However, this increase is canceled by the energetic cost of distorting the DNA from its relaxed conformation. Thus, for *Eco*RV endonuclease, there is little difference in binding affinity for cognate and nonspecific DNA fragments. However, the distortion in the cognate complex dramatically affects catalysis by completing the magnesium ion binding site. *This example illustrates how enzymes can utilize available binding energy to deform substrates and poise them for chemical transformation.* Interactions that take place within the distorted substrate complex stabilize the transition state leading to DNA hydrolysis.

The distortion in the DNA explains how methylation blocks catalysis and protects host-cell DNA. When a methyl group is added to the amino group of the adenine nucleotide at the 5′ end of the recognition sequence, the methyl group's presence precludes the formation of a hydrogen bond between the amino group and the side-chain carbonyl group of asparagine 185. This asparagine residue is closely linked to the other amino acids that form specific contacts with the DNA. The absence of the hydrogen bond disrupts other interactions between the enzyme and the DNA substrate, and the distortion necessary for cleavage will not take place.

TYPE II RESTRICTION ENZYMES HAVE A CATALYTIC CORE IN COMMON AND ARE PROBABLY RELATED BY HORIZONTAL GENE TRANSFER

Type II restriction enzymes are prevalent in Archaea and Eubacteria. What can we tell of the evolutionary history of these enzymes? Comparison of the amino acid sequences of a variety of type II restriction endonucleases did not reveal significant sequence similarity between most pairs of enzymes. However, a careful examination of three-dimensional structures, taking into account the location of the active sites, revealed the presence of a core structure conserved in the different enzymes. This structure includes β strands that contain the aspartate (or, in some cases, glutamate) residues forming the magnesium ion binding sites.

These observations indicate that many type II restriction enzymes are indeed evolutionary related. Analyses of the sequences in greater detail suggest that bacteria may have obtained genes encoding these enzymes from other species by *horizontal gene transfer*, the passing between species of pieces of DNA (such as plasmids) that provide a selective advantage in a particular environment. For example, *Eco*RI (from *E. coli*) and *Rsr*I (from Rhodobacter sphaeroides) are 50per cent identical in sequence over 266 amino acids, clearly indicative of a close evolutionary relationship. However, these species of bacteria are not closely related, as is known from sequence comparisons of other genes and other evidence. Thus, it appears that these species obtained the gene for this restriction endonuclease from a common source more recently than the time of their evolutionary divergence. Moreover, the gene encoding EcoRI endonuclease uses particular codons to specify given amino acids that are strikingly different from the codons used by most E. coli genes, which suggests that the gene did not originate in E. coli. Horizontal gene transfer may be a relatively common event. For example, genes that inactivate antibiotics are often transferred, leading to the transmission of antibiotic resistance from one species to another. For restriction-modification systems, protection against viral infections may have favoured horizontal gene transfer.

NUCLEOSIDE MONOPHOSPHATE KINASES: CATALYZING PHOSPHORYL GROUP EXCHANGE BETWEEN NUCLEOTIDES WITHOUT PROMOTING HYDROLYSIS

The final enzymes that we shall consider are the nucleoside monophosphate kinases (NMP kinases), typified by adenylate kinase. These enzymes catalyze the transfer of the terminal phosphoryl group from a nucleoside triphosphate (NTP), usually ATP, to the phosphoryl group on a nucleoside monophosphate. The challenge for NMP kinases is to promote the transfer of the phosphoryl group from NTP to NMP without promoting the competing reaction—the transfer of a phosphoryl group from NTP to water; that is, NTP hydrolysis. We shall see how the use of induced fit by these enzymes is used to solve this problem. Moreover, these enzymes employ metal ion catalysis; but, in this case, the metal forms a complex with the substrate to enhance enzyme-substrate interaction.

NMP KINASES ARE A FAMILY OF ENZYMES CONTAINING P-LOOP STRUCTURES

X-ray crystallographic methods have yielded the three-dimensional structures of a number of different NMP kinases, both free and bound to substrates or substrate analogs. Comparison of these structures reveals that these enzymes form a family of homologous proteins. In particular, such comparisons reveal the presence of a conserved NTP-binding domain. This domain consists of a central β sheet, surrounded on both sides by α helices. A characteristic feature of this domain is a loop between the first β strand and the first helix. This loop, which typically has an amino acid sequence of the form Gly-X-X-X-X-Gly-Lys, is often referred to as the *P-loop* because it interacts with phosphoryl groups on the bound nucleotide. As described, similar domains containing P-loops are present in a wide variety of important nucleotide-binding proteins.

MAGNESIUM (OR MANGANESE) COMPLEXES OF NUCLEOSIDE TRIPHOSPHATES ARE THE TRUE SUBSTRATES FOR ESSENTIALLY ALL NTP-DEPENDENT ENZYMES

Kinetic studies of NMP kinases, as well as many other enzymes having ATP or other nucleoside triphosphates as a substrate, reveal that these enzymes are essentially inactive in the absence of divalent metal ions such as magnesium (Mg^{2+}) or manganese (Mn^{2+}), but acquire activity on the addition of these ions. In contrast with the enzymes discussed so far, the metal is not a component of the active site. Rather, nucleotides such as ATP bind these ions, and *it is the metal ion-nucleotide complex that is the true substrate for the enzymes.* The dissociation constant for the ATP-Mg^{2+} complex is approximately 0.1 mM, and thus, given that intracellular Mg^{2+} concentrations are typically in the millimolar range, essentially all nucleoside triphosphates are present as NTP-Mg^{2+} complexes.

How does the binding of the magnesium ion to the nucleotide affect catalysis? There are a number of related consequences, but all serve to enhance the specificity of the enzyme-substrate interactions by enhancing binding energy. First, the magnesium ion neutralizes some of the negative charges present on the polyphosphate chain, reducing nonspecific ionic interactions between the enzyme and the polyphosphate group of the nucleotide. Second, the interactions between the magnesium ion and the oxygen atoms in the phosphoryl group hold the nucleotide in well-defined conformations that can be specifically bound by the enzyme. Magnesium ions are typically coordinated to six groups in an octahedral arrangement. Typically, two oxygen atoms are directly coordinated to a magnesium ion, with the remaining coordination positions often occupied by water molecules. Oxygen atoms of the α and β, β and γ, or α and γ phosphoryl groups may contribute, depending on the particular enzyme. In addition, different stereoisomers are produced, depending on exactly which oxygen atoms bind to the metal ion. Third, the magnesium ion provides additional points of interaction between the ATP-Mg^{2+} complex and the enzyme, thus increasing the binding energy. In some cases, such as the DNA polymerases, side chains (often aspartate and glutamate residues) of the enzyme can bind directly to the magnesium ion. In other cases, the enzyme interacts indirectly with the magnesium ion through hydrogen bonds to the coordinated water molecules. Such interactions have been observed in adenylate kinases bound to ATP analogs.

ATP BINDING INDUCES LARGE CONFORMATIONAL CHANGES

Comparison of the structure of adenylate kinase in the presence and absence of an ATP analog reveals that substrate binding induces large structural changes in the kinase, providing a classic example of the use of induced fit. The P-loop closes down on top of the polyphosphate chain, interacting most extensively with the β phosphoryl group. The movement of the P-loop permits the top domain of the enzyme to move down to form a lid over the bound nucleotide. This motion is favoured by interactions between basic residues (conserved among the NMP kinases), the peptide backbone NH groups, and the nucleotide. With the ATP nucleotide held in this position, its γ phosphoryl group is positioned next to the binding site for the second substrate, NMP. In sum, the direct interactions with the nucleotide substrate lead to local structural rearrangements (movement of the P-loop) within the enzyme, which in turn allow more extensive changes (the closing down of the top domain) to take place. The binding of the second substrate, NMP, induces additional conformational changes. Both sets of changes ensure that a catalytically competent conformation is formed only when *both* the donor and the acceptor are bound, preventing wasteful transfer of the phosphoryl group to water. The enzyme holds its two substrates close together and appropriately oriented to stabilize the transition state that leads to the transfer of a phosphoryl group from the ATP to the NMP. This is an example of *catalysis by approximation*. We will see such examples of a catalytically competent active site being generated only on substrate binding many times in our study of biochemistry.

P-LOOP NTPASE DOMAINS ARE PRESENT IN A RANGE OF IMPORTANT PROTEINS

Domains similar (and almost certainly homologous) to those found in NMP kinases are present in a remarkably wide array of proteins, many of which participate in essential biochemical processes. Examples include ATP synthase, the key enzyme responsible for ATP generation; molecular motor proteins such as myosin; signal-transduction proteins such as transducin; proteins essential for translating mRNA into proteins, such as elongation factor Tu; and DNA and RNA unwinding helicases. The wide utility of P-loop NTPase domains is perhaps best explained by their ability to undergo substantial conformational changes on nucleoside triphosphate binding and hydrolysis. We shall encounter these domains (hereafter referred to as P-loop NTPases) throughout the book and shall observe how they function as springs, motors, and clocks. To allow easy recognition of these domains, they, like the binding domains of the NMP kinases, will be depicted with the inner surfaces of the ribbons in a ribbon diagram shown in purple and the P-loop shown in green.

REGULATORY STRATEGIES: ENZYMES AND HEMOGLOBIN

The activity of proteins, including enzymes, often must be regulated so that they function at the proper time and place. The biological activity of proteins is regulated in four principal ways:

1. *Allosteric control.* Allosteric proteins contain distinct regulatory sites and multiple functional sites. Regulation by small signal molecules is a significant means of controlling the activity of many proteins. The binding of these regulatory molecules at sites distinct from the active site triggers conformational changes that are transmitted to the active site. Moreover, allosteric proteins show the property of *cooperativity:* activity at one functional site affects the activity at others. As a consequence, a slight change in substrate concentration can produce substantial changes in activity. Proteins displaying allosteric control are thus information transducers: their activity can be modified in response to signal molecules or to information shared among active sites. This chapter examines two of the best-understood allosteric proteins: the enzyme *aspartate transcarbamoylase* (ATCase) and the oxygen-carrying protein *hemoglobin*. Catalysis by aspartate transcarbamoylase of the first step in pyrimidine biosynthesis is inhibited by cytidine triphosphate, the final product of that biosynthesis, in an example of *feedback inhibition*. The binding of O_2 by hemoglobin is cooperative and is regulated by H^+, CO_2 and 2,3-bisphosphoglycerate (2,3-BPG).

2. *Multiple forms of enzymes.* Isozymes, or isoenzymes, provide an avenue for varying regulation of the same reaction at distinct locations or times. Isozymes are homologous enzymes within a single organism that catalyze the same reaction but differ slightly in structure and more obviously in K_M and V_{max} values, as well as regulatory properties. Often, isozymes are expressed in a distinct tissue or organelle or at a distinct stage of development.

3. *Reversible covalent modification.* The catalytic properties of many enzymes are markedly altered by the covalent attachment of a modifying group, most commonly a phosphoryl group. ATP serves as the phosphoryl donor in these reactions, which are catalyzed by *protein kinases.* The removal of phosphoryl groups by hydrolysis is catalyzed by *protein phosphatases.* This chapter considers the structure, specificity, and control of *protein kinase A* (PKA), a ubiquitous eukaryotic enzyme that regulates diverse target proteins.

4. *Proteolytic activation.* The enzymes controlled by some of these mechanisms cycle between active and inactive states. A different regulatory motif is used to *irreversibly* convert an inactive enzyme into an active one. Many enzymes are activated by the hydrolysis of a few or even one peptide bond in inactive precursors called *zymogens* or *proenzymes.* This regulatory mechanism generates digestive enzymes such as chymotrypsin, trypsin, and pepsin. Caspases, which are proteolytic enzymes that are the executioners in *programmed cell death,* or *apoptosis* are proteolytically activated from the procaspase form. Blood clotting is due to a remarkable cascade of zymogen activations. Active digestive and clotting enzymes are switched off by the irreversible binding of specific inhibitory proteins that are irresistible lures to their molecular prey.

To begin, we will consider the principles of allostery by examining two proteins: the enzyme aspartate transcarbamoylase and the oxygen-transporting protein hemoglobin.

ASPARTATE TRANSCARBAMOYLASE IS ALLOSTERICALLY INHIBITED BY THE END PRODUCT OF ITS PATHWAY

Aspartate transcarbamoylase catalyzes the first step in the biosynthesis of pyrimidines, bases that are components of nucleic acids. The reaction catalyzed by this enzyme is the condensation of aspartate and carbamoyl phosphate to form *N*-carbamoylaspartate and orthophosphate. ATCase catalyzes the committed step in the pathway that will ultimately yield pyrimidine nucleotides such as cytidine triphosphate (CTP). How is this enzyme regulated to generate precisely the amount of CTP needed by the cell?

John Gerhart and Arthur Pardee found that ATCase is inhibited by CTP, the final product of the ATCase-controlled pathway. The rate of the reaction catalyzed by ATCase is fast in the absence of high concentrations of CTP but decreases as the CTP concentration increases. Thus, more molecules are sent along the pathway to make new pyrimidines until sufficient quantities of CTP have accumulated. The effect of CTP on the enzyme exemplifies the *feedback*, or *end-product*, *inhibition* mentioned earlier. Despite the fact that end-product regulation makes considerable physiological sense, the observation that ATCase is inhibited by CTP is remarkable because *CTP is structurally quite different from the substrates of the reaction*. Owing to this structural dissimilarity, CTP must bind to a site distinct from the active site where substrate binds. Such sites are called *allosteric* (from the Greek *allos*, "other," and *stereos*, "structure") or *regulatory sites*. CTP is an example of an *allosteric inhibitor*. In ATCase (but not all allosterically regulated enzymes), the catalytic sites and the regulatory sites are on separate polypeptide chains.

ACTASE CONSISTS OF SEPARABLE CATALYTIC AND REGULATORY SUBUNITS

What is the evidence that ATCase has distinct regulatory and catalytic sites? ATCase can be literally separated into regulatory and catalytic subunits by treatment with a mercurial compound such as *p*-hydroxymercuribenzoate, which reacts with sulfhydryl groups. The results of ultracentrifugation studies carried out by Gerhart and Howard Schachman showed that *p*-hydroxymercuribenzoate dissociates ATCase into two kinds of subunits. The sedimentation coefficient of the native enzyme is 11.6S, whereas those of the dissociated subunits are 2.8S and 5.8S, indicating subunits of different size. The subunits can be readily separated by ion-exchange chromatography because they differ markedly in charge or by centrifugation in a sucrose density gradient because they differ in size. Furthermore, the attached *p*-mercuribenzoate groups can be removed from the separated subunits by adding an excess of mercaptoethanol. The isolated subunits provide materials that can be used to investigate and characterize the individual subunits and their interactions with one another.

The larger subunit is called the *catalytic* (or *c*) *subunit*. This subunit displays catalytic activity, but it is not affected by CTP. The isolated smaller subunit can bind CTP, but has no catalytic activity. Hence, that subunit is called the *regulatory* (or *r*) *subunit*. The catalytic subunit, which consists of three chains (34 kd each), is referred to as c_3. The regulatory subunit, which consists of two chains (17 kd each), is referred to as r_2. The catalytic and regulatory subunits combine rapidly when they are mixed. The resulting complex has the same structure, c_6r_6, as the native enzyme: two catalytic trimers and three regulatory dimers.

$$2c_3 + 3r_2 \rightarrow c_6r_6$$

Furthermore, the reconstituted enzyme has the same allosteric properties as the native enzyme. Thus, ATCase is composed of discrete catalytic and regulatory subunits, which interact in the native enzyme to produce its allosteric behaviour.

ALLOSTERIC INTERACTIONS IN AT CASE ARE MEDIATED BY LARGE CHANGES IN QUATERNARY STRUCTURE

How can the binding of CTP to a regulatory subunit influence reactions at the active site of a catalytic subunit? Significant clues have been provided by the determination of the three-dimensional structure of ATCase in various forms by x-ray crystallography in the laboratory of William Lipscomb. The structure of the enzyme without any ligands bound to it confirms the overall structure of the enzyme. Two catalytic trimers are stacked one on top of the other, linked by three dimers of the regulatory chains. There are significant contacts between the two catalytic trimers: each r chain within a regulatory dimer interacts with a c chain within a catalytic trimer through a structural domain stabilized by a zinc ion bound to four cysteine residues. The ability of *p*-hydroxymercuribenzoate to dissociate the catalytic and regulatory subunits is related to the ability of mercury to bind strongly to the cysteine residues, displacing the zinc and destabilizing this domain.

To understand the mechanism of allosteric regulation, it is crucial to locate each active site and each regulatory site in the three-dimensional structure. To locate the active sites, the enzyme was crystallized in the presence of *N*-(phosphonacetyl)-l-aspartate (PALA), a bisubstrate analog (an analog of the two substrates) that resembles an intermediate along the pathway of catalysis. PALA is a potent competitive inhibitor of ATCase; it binds to and blocks the active sites. The structure of the ATCase-PALA complex reveals that PALA binds at sites lying at the boundaries between pairs of c chains within a catalytic trimer. Note that, though most of the residues belong to one subunit, several key residues belong to a neighbouring subunit. Thus, because the active sites are at the subunit interface, each catalytic trimer contributes three active sites to the complete enzyme. Suitable amino acid residues are available in the active sites for recognizing all features of the bisubstrate analog, including the phosphate and both carboxylate groups.

Further examination of the ATCase-PALA complex reveals a remarkable change in quaternary structure on binding of PALA. The two catalytic trimers move 12 Å farther apart and rotate approximately 10 degrees about their common threefold axis of symmetry. Moreover, the regulatory dimers rotate approximately 15 degrees to accommodate this motion. The enzyme literally expands on PALA binding. In essence, ATCase has two distinct quaternary forms: one that predominates in the absence of substrate or substrate analogs and another that predominates when substrates or analogs are bound. These forms will be referred to as the T (for tense) state and the R (for relaxed) state, respectively. The T state has lower affinity for substrates and, hence, lower catalytic activity than does the R state. In the presence of any fixed concentration of aspartate and carbamoyl phosphate, the enzyme exists in equilibrium between the T and the R forms. *The position of the equilibrium depends on the number of active sites that are occupied by substrate.*

Having located the active sites and seen that PALA binding results in substantial structural changes in the entire ATCase molecule, we now turn our attention to the effects of CTP. Where on the regulatory subunit does CTP bind? Determination of the structure of ATCase in the presence of CTP reveals a binding site for this nucleotide in each regulatory chain in a domain that does not interact with the catalytic subunit. The question naturally arises as to how CTP can inhibit the catalytic activity of the enzyme when it does not interact with the catalytic chain. Each active site is more than 50 Å from the nearest CTP binding site. The CTP-bound form is in the T quaternary state in the absence of bound substrate.

The quaternary structural changes observed on substrate-analog binding suggest a mechanism for the allosteric regulation of ATCase by CTP. The binding of the inhibitor CTP shifts the equilibrium toward the T state, decreasing the net enzyme activity and reducing the rate of *N*-carbamoylaspartate generation. This mechanism for allosteric regulation is referred to as the *concerted mechanism* because the change in the enzyme is "all or none"; the entire enzyme is converted from T into R, affecting all of the catalytic sites equally. The concerted mechanism stands in contrast with the sequential mechanism, which will be discussed shortly.

ALLOSTERICALLY REGULATED ENZYMES DO NOT FOLLOW MICHAELIS-MENTEN KINETICS

Allosteric enzymes are distinguished by their response to substrate concentration in addition to their susceptibility to regulation by other molecules. Examining the rate of product formation as a function of substrate concentration can be a source of further insights into the mechanism of regulation of ATCase. The curve differs from that expected for an enzyme that follows Michaelis-Menten kinetics. The observed curve is referred to as sigmoid because it resembles an "S." How can we explain this kinetic behaviour in light of the structural observations? In the absence of substrate, the enzyme exists almost entirely in the T state. However, the binding of substrate molecules to the enzyme shifts the enzyme toward the R state. A transition from T to R favoured by substrate binding to one site will increase the enzymatic activity of the remaining five sites, leading to an overall increase in enzyme activity. This important property is called *cooperativity* because the subunits cooperate with one another. If one subunit switches conformation, they all do. The sigmoid curve can be pictured as a composite of two Michaelis-Menten curves, one corresponding to the T state and the other to the R state. An increase in substrate concentration favours a transition from the T-state curve to the R-state curve.

The importance of the changes in quaternary structure in determining the sigmoidal curve is illustrated nicely by studies of the isolated catalytic trimer, freed by *p*-hydroxymercuribenzoate treatment. The catalytic subunit shows Michaelis-Menten kinetics with kinetic parameters that are indistinguishable from those deduced for the R state. Thus, the term *tense* is apt: in the T state, the regulatory dimers hold the two catalytic trimers sufficiently close to one another that key loops on their surfaces collide and interfere with conformational adjustments necessary for high-affinity substrate binding and catalysis.

ALLOSTERIC REGULATORS MODULATE THE T-TO-R EQUILIBRIUM

What is the effect of CTP on the kinetic profile of ATCase? CTP increases the initial phase of the sigmoidal curve. As noted earlier, CTP inhibits the activity of ATCase. In the presence of CTP, the enzyme becomes less responsive to the cooperative effects facilitated by substrate binding; more substrate is required to attain a given reaction rate. Interestingly, ATP, too, is an allosteric effector of ATCase. However, the effect of ATP is to *increase* the reaction rate at a given aspartate concentration. At high concentrations of ATP, the kinetic profile shows a lesspronounced sigmoidal behaviour. Note that such sigmoidal behaviour has an additional consequence: in the concentration range where the T-to-R transition is taking place, the curve depends quite steeply on the substrate concentration. The effects of substrates on allosteric enzymes are referred to as *homotropic effects* (from the Greek *homós,* "same"). In contrast, the effects of nonsubstrate molecules on allosteric enzymes (such as those of CTP and ATP on ATCase) are referred to as *heterotropic effects* (from the Greek *héteros,* "different").

The increase in ATCase activity in response to increased ATP concentration has two potential physiological explanations. First, high ATP concentration signals a high concentration of purine nucleotides in the cell; the increase in ATCase activity will tend to balance the purine and pyrimidine pools. Second, a high concentration of ATP indicates that there is significant energy stored in the cell to promote mRNA synthesis and DNA replication.

THE CONCERTED MODEL CAN BE FORMULATED IN QUANTITATIVE TERMS

The concerted model was first proposed by Jacques Monod, Jeffries Wyman, and Jean-Pierre Changeux; hence, it is often referred to as the MWC model. This model can be formulated in quantitative terms. Consider an enzyme with n identical active sites. Suppose that the enzyme exists in equilibrium between a T form with a low affinity for its substrate and an R form with a high affinity for the substrate. We can define L as the equilibrium constant between the R and the T forms; c as the ratio of the affinities of the two forms for the substrate, S, measured as dissociation constants; and α as the ratio of substrate concentration to the dissociation constant K_R.

$$\mathrm{R} \rightleftharpoons \mathrm{T} \qquad L = \frac{[\mathrm{T}]}{[\mathrm{R}]}$$

$$\mathrm{Site_R} + \mathrm{S} \rightleftharpoons \mathrm{Site_R} - \mathrm{S} \qquad K_R = \frac{[\mathrm{Site_R}][\mathrm{S}]}{[\mathrm{Site_R} - \mathrm{S}]}$$

$$\mathrm{Site_T} + \mathrm{S} \rightleftharpoons \mathrm{Site_T} - \mathrm{S} \qquad K_T = \frac{[\mathrm{Site_T}][\mathrm{S}]}{[\mathrm{Site_T} - \mathrm{S}]}$$

Define

$$c = \frac{K_R}{K_T} \quad \text{and} \quad \alpha = \frac{[\mathrm{S}]}{K_R}$$

The fraction of active sites bound to substrate (fractional saturation, Y_S) is given by

$$Y_S = \frac{\alpha(1-\alpha)^{n-1} + Lc\alpha(1+c\alpha)^{n-1}}{(1+\alpha)^n + L(1+c\alpha)^n}$$

where n is the number of sites in the enzyme.

This quantitative model can be used to examine the data from ATCase, for which $n = 6$. Excellent agreement with experimental data is obtained with L M" 200 and c M" 0.1. Thus, in the absence of bound substrate, the equilibrium favours the T form by a factor of 200 (i.e., only 1 in 200 molecules is in the R form), and the affinity of the R form for substrate is approximately 10 times as high as that of the T form. As substrate binds to each active site, the equilibrium shifts toward the R form. For example, with these parameters, when half the active sites (three of six) are occupied by substrate, the equilibrium has shifted so that the ratio of T to R is now 1 to 5; that is, nearly all the molecules are in the R form.

The effects of CTP and ATP can be modeled simply by changing the value of *L*. For the CTP-saturated form, the value of *L* increases to 1250. Thus, it takes more substrate to shift the equilibrium appreciably to the R form. For the ATP saturated form, the value of *L* decreases to 70.

SEQUENTIAL MODELS ALSO CAN ACCOUNT FOR ALLOSTERIC EFFECTS

In the concerted model, an allosteric enzyme can exist in one of only two states, T and R; no intermediate states are allowed. An alternative, first proposed by Daniel Koshland, posits that *sequential* changes in structure take place within an oligomeric enzyme as active sites are occupied. The binding of substrate to one site influences the substrate affinity of neighbouring active sites *without necessarily inducing a transition encompassing the entire enzyme*. An important feature of sequential in contrast with concerted models is that the former can account for *negative cooperativity*, in which the binding of substrate to one active site *decreases* the affinity of other sites for substrate. The results of studies of a number of allosteric proteins suggest that most behave according to some combination of the sequential and cooperative models.

HEMOGLOBIN TRANSPORTS OXYGEN EFFICIENTLY BY BINDING OXYGEN COOPERATIVELY

Allostery is a property not limited to enzymes. The basic principles of allostery are also well illustrated by the oxygen-transport protein hemoglobin. The binding of oxygen to hemoglobin isolated from red blood cells displays marked sigmoidal behaviour (similar to that observed for the activity of ATCase, as a function of substrate concentration), which is indicative of cooperation between subunits. What is the physiological significance of the cooperative binding of oxygen by hemoglobin? Oxygen must be transported in the blood from the lungs, where the partial pressure of oxygen (pO_2) is relatively high (approximately 100 torr), to the tissues, where the partial pressure of oxygen is much lower (typically 20 torr). Let us consider how the cooperative behaviour represented by the sigmoidal curve leads to efficient oxygen transport. In the lungs, hemoglobin becomes nearly saturated with oxygen such that 98per cent of the oxygen-binding sites are occupied. When hemoglobin moves to the tissues, the saturation level drops to 32per cent. Thus, a total of 98 - 32 = 66per cent of the potential oxygen-binding sites contribute to oxygen transport. In comparison, for a hypothetical noncooperative transport protein, the most oxygen that can be transported from a region in which pO_2 is 100 torr to one in which it is 20 torr is 63 - 25 = 38per cent. Thus, *the cooperative binding of oxygen by hemoglobin enables it to deliver 1.7 times as much oxygen as it would if the sites were independent*. The homotropic regulation of hemoglobin by its ligand oxygen dramatically increases its physiological oxygen-carrying capacity.

OXYGEN BINDING INDUCES SUBSTANTIAL STRUCTURAL CHANGES AT THE IRON SITES IN HEMOGLOBIN

The results of structural studies pioneered by Max Perutz revealed the structure of hemoglobin in various forms. Human hemoglobin A, present in adults, consists of four subunits: two α subunits and two β subunits. The α and β subunits are homologous and have similar three-dimensional structures. The capacity of hemoglobin to bind oxygen depends on the presence of a bound prosthetic group called *heme*. The heme group is responsible for the distinctive red colour of blood. The heme group consists of an organic component and a central iron atom. The organic component, called *protoporphyrin*, is made up of four pyrrole rings linked by methene bridges to form a tetrapyrrole ring. Four methyl groups, two vinyl groups, and two propionate side chains are attached.

Heme
(Fe-protoporphyrin IX)

The iron atom lies in the centre of the protoporphyrin, bonded to the four pyrrole nitrogen atoms. Under normal conditions, the iron is in the ferrous (Fe^{2+}) oxidation state. The iron ion can form two additional bonds, one on each side of the heme plane. These binding sites are called the fifth and sixth coordination sites. In hemoglobin, the fifth coordination site is occupied by the imidazole ring of a histidine residue from the protein. In deoxyhemoglobin, the sixth coordination site remains unoccupied. The iron ion lies approximately 0.4 Å outside the porphyrin plane because iron, in this form, is slightly too large to fit into the well-defined hole within the porphyrin ring.

The binding of the oxygen molecule at the sixth coordination site of the iron ion substantially rearranges the electrons within the iron so that the ion becomes effectively smaller, allowing it to move into the plane of the porphyrin.

This change in electronic structure is paralleled by changes in the magnetic properties of hemoglobin, which are the basis for functional magnetic resonance imaging. Indeed, the structural changes that take place on oxygen binding were anticipated by Linus Pauling, based on magnetic measurements in 1936, nearly 25 years before the three-dimensional structure of hemoglobin was elucidated.

OXYGEN BINDING MARKEDLY CHANGES THE QUATERNARY STRUCTURE OF HEMOGLOBIN

The three-dimensional structure of hemoglobin is best described as a pair of identical αβ dimers ($\alpha_1\beta_1$ and $\alpha_2\beta_2$) that associate to form the hemoglobin tetramer. In deoxyhemoglobin, these αβ dimers are linked by an extensive interface, which includes, among other regions, the carboxyl terminus of each chain. The heme groups are well separated in the tetramer with iron-iron distances ranging from 24 to 40 Å. The deoxy form corresponds to the T state in the context of either the concerted or the sequential model for hemoglobin cooperativity. On oxygen binding, there are substantial changes in quaternary structure that correspond to the T-to-R state transition.

The $\alpha_1\beta_1$ and $\alpha_2\beta_2$ dimers rotate approximately 15 degrees with respect to one another. The dimers themselves are relatively unchanged, although localized conformational shifts do occur. Thus, the interface between the $\alpha_1\beta_1$ and $\alpha_2\beta_2$ dimers is most effected by this structural transition.

How does oxygen binding lead to the structural transition from the T state to the R state? When the iron ion moves into the plane of the porphyrin, the histidine residue bound in the fifth coordination site moves with it. This histidine residue is part of an α helix, which also moves. The carboxyl terminal end of this α helix lies in the interface between the two αβ dimers. Consequently, the structural transition at the iron ion is directly transmitted to the other subunits. The rearrangement of the dimer interface provides a pathway for communication between subunits, enabling the cooperative binding of oxygen.

Is the cooperative binding of oxygen by hemoglobin best described by the concerted or the sequential model? Neither model in its pure form fully accounts for the behaviour of hemoglobin. Instead, a combined model is required. Hemoglobin behaviour is concerted in that hemoglobin with three sites occupied by oxygen is in the quaternary structure associated with the R state. The remaining open binding site has an affinity for oxygen more than 20-fold as great as that of fully deoxygenated hemoglobin binding its first oxygen. However, the behaviour is not fully concerted, because hemoglobin with oxygen bound to only one of four sites remains in the T-state quaternary structure. Yet, this molecule binds oxygen 3 times as strongly as does fully deoxygenated hemoglobin, an observation consistent only with a sequential model. These results highlight the fact that the concerted and sequential models represent idealized limiting cases, which real systems may approach but rarely attain.

TUNING THE OXYGEN AFFINITY OF HEMOGLOBIN: THE EFFECT OF 2,3-BISPHOSPHOGLYCERATE

Examination of the oxygen binding of hemoglobin fully purified from red blood cells revealed that the oxygen affinity of purified hemoglobin is much greater than that for hemoglobin within red blood cells. This dramatic difference is due to the presence within these cells of 2,3-bisphosphoglycerate (2,3-BPG) (also known as 2,3-diphosphoglycerate or 2,3-DPG). This highly anionic compound is present in red blood cells at approximately the same concentration as that of hemoglobin (~ 2 mM). Without 2,3-BPG, hemoglobin would be an extremely inefficient oxygen transporter, releasing only 8per cent of its cargo in the tissues.

How does 2,3-BPG affect oxygen affinity so significantly? Examination of the crystal structure of deoxyhemoglobin in the presence of 2,3-BPG reveals that a single molecule of 2,3-BPG binds in a pocket, present only in the T form, in the centre of the hemoglobin tetramer. On T-to-R transition, this pocket collapses. Thus, 2,3-BPG binds preferentially to deoxyhemoglobin and stabilizes it, effectively reducing the oxygen affinity. In order for the structural transition from T to R to take place, the bonds between hemoglobin and 2,3-BPG must be broken and 2,3-BPG must be expelled.

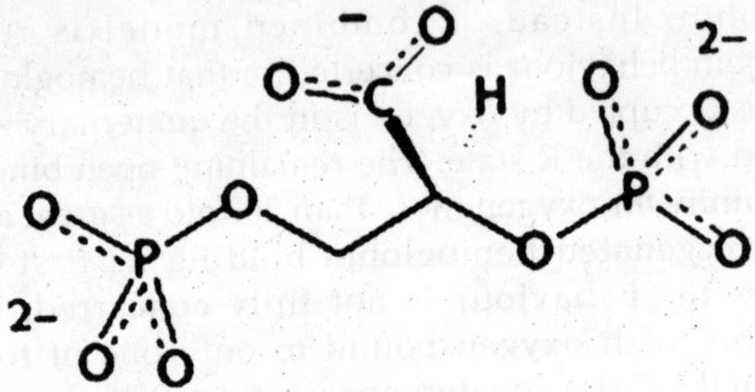

2,3-Bisphosphoglycerate
(2,3-BPG)

2,3-BPG binding to hemoglobin has other crucial physiological consequences. The globin gene expressed by fetuses differs from that expressed by human adults; fetal hemoglobin tetramers include two α chains and two γ chains. The γ chain, a result of another gene duplication, is 72per cent identical in amino acid sequence with the β chain. One noteworthy change is the substitution of a serine residue for His 143 in the β chain of the 2,3-BPG-binding site. This change removes two positive charges from the 2,3-BPG-binding site (one from each chain) and reduces the affinity of 2,3-BPG for fetal hemoglobin, thereby increasing the oxygen-binding affinity of fetal hemoglobin relative to that of maternal (adult) hemoglobin. This difference in oxygen affinity allows oxygen to be effectively transferred from maternal to fetal red cells. We see again an example of where gene duplication and specialization produced a ready solution to a biological challenge—in this case, the transport of oxygen from mother to fetus.

THE BOHR EFFECT: HYDROGEN IONS AND CARBON DIOXIDE PROMOTE THE RELEASE OF OXYGEN

Rapidly metabolizing tissues, such as contracting muscle, have a high need for oxygen and generate large amounts of hydrogen ions and carbon dioxide as well. Both of these species are heterotropic effectors of hemoglobin that enhance oxygen release. The oxygen affinity of hemoglobin decreases as pH decreases from the value of 7.4 found in the lungs. Thus, as hemoglobin moves into a region of low pH, its tendency to release oxygen increases. For example, transport from the lungs, with pH 7.4 and an oxygen partial pressure of 100 torr, to active muscle, with a pH of 7.2 and an oxygen partial pressure of 20 torr, results in a release of oxygen amounting to 77per cent of total carrying capacity. Recall that only 66per cent of the oxygen would be released in the absence of any change in pH. In addition, hemoglobin responds to carbon dioxide with a decrease in oxygen affinity, thus facilitating the release of oxygen in tissues with a high carbon dioxide concentration. In the presence of carbon dioxide at a partial pressure of 40 torr, the amount of oxygen released approaches 90per cent of the maximum carrying capacity. Thus, the heterotropic regulation of hemoglobin by hydrogen ions and carbon dioxide further increases the oxygen-transporting efficiency of this magnificent allosteric protein.

The regulation of oxygen binding by hydrogen ions and carbon dioxide is called the *Bohr effect* after Christian Bohr, who described this phenomenon in 1904. The results of structural and chemical studies have revealed much about the chemical basis of the Bohr effect. At least two sets of chemical groups are responsible for the effect of protons: the amino termini and the side chains of histidines β146 and α122, which have pK_a values near pH 7. Consider histidine β146. In deoxyhemoglobin, the terminal carboxylate group of β146 forms a salt bridge with a lysine residue in the α subunit of the other αβ dimer. This interaction locks the side chain of histidine β146 in a position where it can participate in a salt bridge with negatively charged aspartate 94 in the same chain, provided that the imidazole group of the histidine residue is protonated. At high pH, the side chain of histidine β146 is not protonated and the salt bridge does not form. As the pH drops, however, the side chain of histidine β146 becomes protonated, the salt bridge with aspartate β94 forms, and the quaternary structure characteristic of deoxyhemoglobin is stabilized, leading to a greater tendency for oxygen to be released at actively metabolizing tissues. No significant change takes place in oxyhemoglobin over the same pH range.

Carbon dioxide also stabilizes deoxyhemoglobin by reacting with the terminal amino groups to form carbamate groups, which are negatively charged, in contrast with the neutral or positive charges on the free amino groups. The amino termini lie at the interface between the αβ dimers, and these negatively charged carbamate groups participate in salt-bridge interactions, characteristic of the T-state structure, which stabilize deoxyhemoglobin's structure and favour the release of oxygen.

$$R(H)N{-}H + O{=}C{=}O \rightleftharpoons R(H)N{-}COO^{-} + H^{+}$$

Carbomate

Hemoglobin with bound carbon dioxide and hydrogen ions is carried in the blood back to the lungs, where it releases the hydrogen ions and carbon dioxide and rebinds oxygen. Thus, hemoglobin helps to transport hydrogen ions and carbon dioxide in addition to transporting oxygen. However, transport by hemoglobin accounts for only about 14per cent of the total transport of these species; both hydrogen ions and carbon dioxide are also transported in the blood as bicarbonate (HCO_3^-) formed spontaneously or through the action of carbonic anhydrase an abundant enzyme in red blood cells.

ISOZYMES PROVIDE A MEANS OF REGULATION SPECIFIC TO DISTINCT TISSUES AND DEVELOPMENTAL STAGES

Isozymes or *isoenzymes*, are enzymes that differ in amino acid sequence yet catalyze the same reaction. Usually, these enzymes display different kinetic parameters, such as K_M, or different regulatory properties. They are encoded by different genetic loci, which usually arise through gene duplication and divergence. Isozymes differ from allozymes, which are enzymes that arise from allelic variation at one gene locus. Isozymes can often be distinguished from one another by biochemical properties such as electrophoretic mobility.

The existence of isozymes permits the fine-tuning of metabolism to meet the particular needs of a given tissue or developmental stage. Consider the example of lactate dehydrogenase (LDH), an enzyme that functions in anaerobic glucose metabolism and glucose synthesis. Human beings have two isozymic polypeptide chains for this enzyme: the H isozyme highly expressed in heart and the M isozyme found in skeletal muscle. The amino acid sequences are 75per cent identical. The functional enzyme is tetrameric, and many different combinations of the two subunits are possible. The H_4 isozyme, found in the heart, has a higher affinity for substrates than does the M_4 isozyme. The two isozymes also differ in that high levels of pyruvate allosterically inhibit the H_4 but not the M_4 isozyme. The other combinations, such as H_3M, have intermediate properties depending on the ratio of the two kinds of chains.

The M_4 isozyme functions optimally in an anaerobic environment, whereas the H_4 isozyme does so in an aerobic environment. Indeed, the proportions of these isozymes are altered in the course of development of the rat heart as the tissue switches from an anaerobic environment to an aerobic one. Figure shows the tissue-specific forms of lactate dehydrogenase in adult rat tissues. The appearance of some isozymes in the blood is indicative of tissue damage and can be used for clinical diagnosis. For instance, an increase in serum levels of H_4 relative to H_3M is an indication that a myocardial infarction, or heart attack, has occurred.

COVALENT MODIFICATION IS A MEANS OF REGULATING ENZYME ACTIVITY

The covalent attachment of another molecule can modify the activity of enzymes and many other proteins. In these instances, a donor molecule provides a functional moiety that modifies the properties of the enzyme. Most modifications are reversible. Phosphorylation and dephosphorylation are the most common but not the only means of covalent modification. Histones—proteins that assist in the packaging of DNA into chromosomes as well as in gene regulation—are rapidly acetylated and deacetylated in vivo. More heavily acetylated histones are associated with genes that are being actively transcribed. The acetyltransferase and deacetylase enzymes are themselves regulated by phosphorylation, showing that the covalent modification of histones may be controlled by the covalent modification of the modifying enzymes.

Modification is not readily reversible in some cases. Some proteins in signal-transduction pathways, such as Ras and Src (a protein tyrosine kinase), are localized to the cytoplasmic face of the plasma membrane by the irreversible attachment of a lipid group. Fixed in this location, the proteins are better able to receive and transmit information that is being passed along their signaling pathways. The attachment of ubiquitin, a protein comprising 72 amino acids, is a signal that a protein is to be destroyed, the ultimate means of regulation. Cyclin, an important protein in cell-cycle regulation, must be ubiquitinated and destroyed before a cell can enter anaphase and proceed through the cell cycle.

Virtually all the metabolic processes that we will examine are regulated in part by covalent modification. Indeed, the allosteric properties of many enzymes are modified by covalent modification.

PHOSPHORYLATION IS A HIGHLY EFFECTIVE MEANS OF REGULATING THE ACTIVITIES OF TARGET PROTEINS

The activities of many enzymes, membrane channels, and other target proteins are regulated by phosphorylation, the most prevalent reversible covalent modification. Indeed, we will see this regulatory mechanism in virtually every metabolic process in eukaryotic cells. The enzymes catalyzing phosphorylation reactions are called *protein kinases*, which constitute one of the largest protein families known, with more than 100 homologous enzymes in yeast and more than 550 in human beings. This multiplicity of enzymes allows regulation to be fine-tuned according to a specific tissue, time, or substrate.

The terminal (γ) phosphoryl group of ATP is transferred to specific *serine* and *threonine* residues by one class of protein kinases and to specific *tyrosine* residues by another.

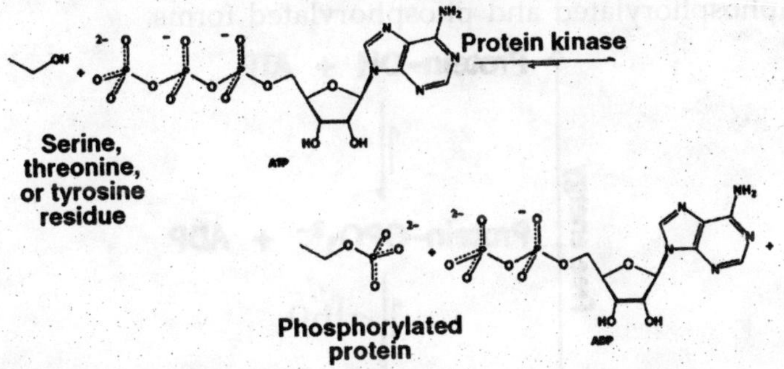

Protein kinase

Serine, threonine, or tyrosine residue

ATP

Phosphorylated protein

ADP

The acceptors in protein phosphorylation reactions are located inside cells, where the phosphoryl-group donor ATP is abundant. Proteins that are entirely extracellular are not regulated by reversible phosphorylation. *Protein phosphatases* reverse the effects of kinases by catalyzing the hydrolytic removal of phosphoryl groups attached to proteins.

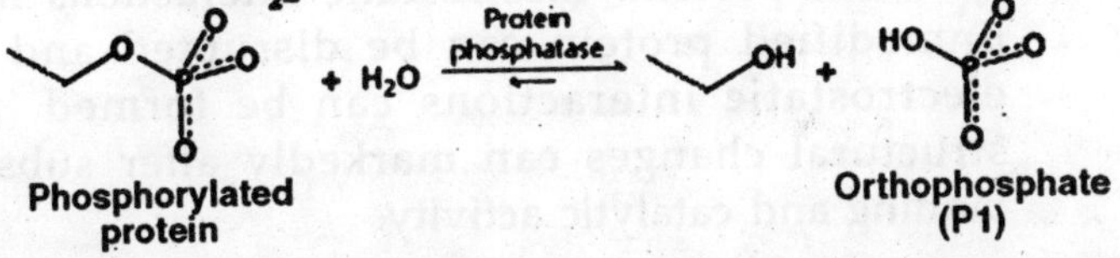

Phosphorylated protein + H_2O —Protein phosphatase→ + Orthophosphate (P1)

The unmodified hydroxyl-containing side chain is regenerated and orthophosphate (P_i) is produced.

It is important to note that phosphorylation and dephosphorylation are not the reverse of one another; each is essentially irreversible under physiological conditions. Furthermore, both reactions take place at negligible rates in the absence of enzymes. Thus, phosphorylation of a protein substrate will take place only through the action of a specific protein kinase and at the expense of ATP cleavage, and dephosphorylation will result only through the action of a phosphatase. The rate of cycling between the phosphorylated and the dephosphorylated states depends on the relative activities of kinases and phosphatases. Note that the net outcome of the two reactions is the hydrolysis of ATP to ADP and P_i, which has a ΔG of -12 kcal mol^{-1} (-50 kJ mol^{-1}) under cellular conditions. This highly favourable free-energy change ensures that target proteins cycle unidirectionally between unphosphorylated and phosphorylated forms.

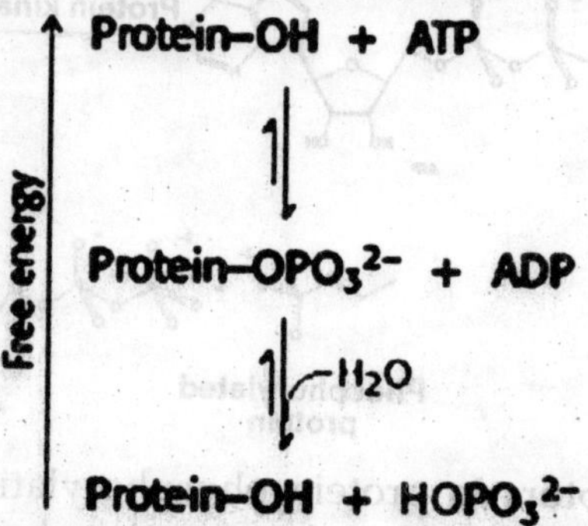

Phosphorylation is a highly effective means of controlling the activity of proteins for structural, thermodynamic, kinetic, and regulatory reasons:

1. A phosphoryl group adds two negative charges to a modified protein. Electrostatic interactions in the unmodified protein can be disrupted and new electrostatic interactions can be formed. Such structural changes can markedly alter substrate binding and catalytic activity.
2. A phosphate group can form three or more hydrogen bonds. The tetrahedral geometry of the phosphoryl group makes these hydrogen bonds highly directional, allowing for specific interactions with hydrogen-bond donors.
3. The free energy of phosphorylation is large. Of the -12 kcal mol^{-1} (-50 kJ mol^{-1}) provided by ATP, about half is consumed in making phosphorylation

irreversible; the other half is conserved in the phosphorylated protein. Recall that a free-energy change of 1.36 kcal mol^{-1} (5.69 kJ mol^{-1}) corresponds to a factor of 10 in an equilibrium constant. Hence, phosphorylation can change the conformational equilibrium between different functional states by a large factor, of the order of 10^4.

4. Phosphorylation and dephosphorylation can take place in less than a second or over a span of hours. The kinetics can be adjusted to meet the timing needs of a physiological process.
5. Phosphorylation often evokes *highly amplified* effects. A single activated kinase can phosphorylate hundreds of target proteins in a short interval. Further amplification can take place because the target proteins may be enzymes, each of which can then transform a large number of substrate molecules.
6. ATP is the cellular energy currency. The use of this compound as a phosphoryl group donor links the energy status of the cell to the regulation of metabolism.

Protein kinases vary in their degree of specificity. *Dedicated protein kinases* phosphorylate a single protein or several closely related ones. *Multifunctional protein kinases* modify many different targets; they have a wide reach and can coordinate diverse processes. Comparisons of amino acid sequences of many phosphorylation sites show that a multifunctional kinase recognizes related sequences. For example, the *consensus sequence* recognized by protein kinase A is Arg-Arg-X-*Ser*-Z or Arg-Arg-X-*Thr*-Z, in which X is a small residue, Z is a large hydrophobic one, and *Ser* or *Thr* is the site of phosphorylation. It should be noted that this sequence is not absolutely required. Lysine, for example, can substitute for one of the arginine residues but with some loss of affinity. Short synthetic peptides containing a consensus motif are nearly always phosphorylated by serine- threonine protein kinases. Thus, *the primary determinant of specificity is the amino acid sequence surrounding the serine or threonine phosphorylation site.* However, distant residues can contribute to specificity. For instance, changes in protein conformation may alter the accessibility of a possible phosphorylation site.

CYCLIC AMP ACTIVATES PROTEIN KINASE A BY ALTERING THE QUATERNARY STRUCTURE

Protein kinases modulate the activity of many proteins, but what leads to the activation of a kinase? Activation is often a multistep process initiated by hormones. In some cases, the hormones trigger the formation of cyclic AMP, a molecule formed by cyclization of ATP. Cyclic AMP serves as an intracellular messenger in mediating the physiological actions of hormones. The striking finding is that *most effects of cAMP in eukaryotic cells are achieved through the activation by cAMP of a single protein kinase. This key enzyme is called protein kinase A or PKA.* The kinase alters the activities of target proteins by phosphorylating specific serine or threonine residues. As we shall see, PKA provides a clear example of the integration of allosteric regulation and phosphorylation.

PKA is activated by cAMP concentrations of the order of 10 nM. The activation mechanism is reminiscent of that of aspartate transcarbamoylase. Like that enzyme, PKA in muscle consists of two kinds of subunits: a 49-kd regulatory (R) subunit, which has high affinity for cAMP, and a 38-kd catalytic (C) subunit. In the absence of cAMP, the regulatory and catalytic subunits form an R_2C_2 complex that is enzymatically inactive. The binding of two molecules of cAMP to each of the regulatory subunits leads to the dissociation of R_2C_2 into an R_2 subunit and two C subunits. These free catalytic subunits are then enzymatically active. Thus, *the binding of cAMP to the regulatory subunit relieves its inhibition of the catalytic subunit.* PKA and most other kinases exist in isozymic forms for finetuning regulation to meet the needs of a specific cell or developmental stage.

How does the binding of cAMP activate the kinase? Each R chain contains the sequence Arg-Arg-Gly-*Ala*-Ile, which matches the consensus sequence for phosphorylation except for the presence of alanine in place of serine. In the R_2C_2 complex, this *pseudosubstrate sequence* of R occupies the catalytic site of C, thereby preventing the entry of protein substrates. The binding of cAMP to the R chains allosterically moves the pseudosubstrate sequences out of the catalytic sites. The released C chains are then free to bind and phosphorylate substrate proteins.

ATP AND THE TARGET PROTEIN BIND TO A DEEP CLEFT IN THE CATALYTIC SUBUNIT OF PROTEIN KINASE A

The three-dimensional structure of the catalytic subunit of PKA containing a bound 20-residue peptide inhibitor was determined by x-ray crystallography. The 350-residue catalytic subunit has two lobes. ATP and part of the inhibitor fill a deep cleft between the lobes. The smaller lobe makes many contacts with ATP-Mg^{2+}, whereas the larger lobe binds peptide and contributes the key catalytic residues. Like other kinases the two lobes move closer to one another on substrate binding; mechanisms for restricting this domain closure provides a means for regulating protein kinase activity. *The PKA structure has broad significance because residues 40 to 280 constitute a conserved catalytic core that is common to essentially all known protein kinases.* We see here an example of a successful biochemical solution to the problem of protein phosphorylation being employed many times in the course of evolution.

The bound peptide in this crystal occupies the active site because it contains the pseudosubstrate sequence Arg-Arg-Asn-*Ala*-Ile. The structure of the complex reveals the basis for the consensus sequence. The guanidinium group of the first arginine residue forms an ion pair with the carboxylate side chain of a glutamate residue (Glu 127) of the enzyme. The second arginine likewise interacts with two other carboxylates. The nonpolar side chain of isoleucine, which matches Z in the consensus sequence, fits snugly in a hydrophobic groove formed by two leucine residues of the enzyme.

MANY ENZYMES ARE ACTIVATED BY SPECIFIC PROTEOLYTIC CLEAVAGE

We turn now to a different mechanism of enzyme regulation. Many enzymes acquire full enzymatic activity as they spontaneously fold into their characteristic three-dimensional forms. In contrast, other enzymes are synthesized as inactive precursors that are subsequently activated by cleavage of one or a few specific peptide bonds. The inactive precursor is called a *zymogen* (or a *proenzyme*). A energy source (ATP) is not needed for cleavage. Therefore, in contrast with reversible regulation by phosphorylation, even proteins located outside cells can be activated by this means. Another noteworthy difference is that proteolytic activation, in contrast with allosteric control and reversible covalent modification, occurs just once in the life of an enzyme molecule.

Specific proteolysis is a common means of activating enzymes and other proteins in biological systems. For example:

1. The *digestive enzymes* that hydrolyze proteins are synthesized as zymogens in the stomach and pancreas.
2. *Blood clotting* is mediated by a cascade of proteolytic activations that ensures a rapid and amplified response to trauma.
3. Some protein hormones are synthesized as inactive precursors. For example, ***insulin*** is derived from *proinsulin* by proteolytic removal of a peptide.
4. The fibrous protein *collagen,* the major constituent of skin and bone, is derived from *procollagen,* a soluble precursor.
5. Many *developmental processes* are controlled by the activation of zymogens. For example, in the metamorphosis of a tadpole into a frog, large amounts of collagen are resorbed from the tail in the course of a few days. Likewise, much collagen is broken down in a mammalian uterus after delivery.

The conversion of *procollagenase* into *collagenase,* the active protease, is precisely timed in these remodeling processes.

6. *Programmed cell death,* or *apoptosis,* is mediated by proteolytic enzymes called *caspases,* which are synthesized in precursor form as *procaspases.* When activated by various signals, caspases function to cause cell death in most organisms, ranging from *C. elegans* to human beings. Apoptosis provides a means sculpting the shapes of body parts in the course of development and a means of eliminating cells producing anti-self antibodies or infected with pathogens as well as cells containing large amounts of damaged DNA.

We next examine the activation and control of zymogens, using as examples several digestive enzymes as well as blood-clot formation.

CHYMOTRYPSINOGEN IS ACTIVATED BY SPECIFIC CLEAVAGE OF A SINGLE PEPTIDE BOND

Chymotrypsin is a digestive enzyme that hydrolyzes proteins in the small intestine. Its inactive precursor, *chymotrypsinogen,* is synthesized in the pancreas, as are several other zymogens and digestive enzymes. Indeed, the pancreas is one of the most active organs in synthesizing and secreting proteins. The enzymes and zymogens are synthesized in the acinar cells of the pancreas and stored inside membrane-bounded granules. The zymogen granules accumulate at the apex of the acinar cell; when the cell is stimulated by a hormonal signal or a nerve impulse, the contents of the granules are released into a duct leading into the duodenum.

Chymotrypsinogen, a single polypeptide chain consisting of 245 amino acid residues, is virtually devoid of enzymatic activity. It is converted into a fully active enzyme when the peptide bond joining arginine 15 and isoleucine 16 is cleaved by trypsin. The resulting active enzyme, called p-chymotrypsin, then acts on other p-chymotrypsin molecules.

Two dipeptides are removed to yield α-chymotrypsin, the stable form of the enzyme. The three resulting chains in α-chymotrypsin remain linked to one another by two interchain disulfide bonds. The striking feature of this activation process is that *cleavage of a single specific peptide bond transforms the protein from a catalytically inactive form into one that is fully active.*

PROTEOLYTIC ACTIVATION OF CHYMOTRYPSINOGEN LEADS TO THE FORMATION OF A SUBSTRATE-BINDING SITE

How does cleavage of a single peptide bond activate the zymogen? Key conformational changes, which were revealed by the elucidation of the three-dimensional structure of chymotrypsinogen, result from the cleavage of the peptide bond between amino acids 15 and 16.

1. The newly formed *amino-terminal group of isoleucine 16 turns inward and forms an ionic bond with aspartate 194* in the interior of the chymotrypsin molecule. Protonation of this amino group stabilizes the active form of chymotrypsin.
2. This electrostatic interaction triggers a number of conformational changes. Methionine 192 moves from a deeply buried position in the zymogen to the surface of the active enzyme, and residues 187 and 193 become more extended. These changes result in the formation of the *substrate-specificity site* for aromatic and bulky nonpolar groups. One side of this site is made up of residues 189 through 192. *This cavity for binding part of the substrate is not fully formed in the zymogen.*
3. The tetrahedral transition state in catalysis by chymotrypsin is stabilized by hydrogen bonds between the negatively charged carbonyl oxygen atom of the substrate and two NH groups of the main chain of the enzyme. One of these NH groups is not appropriately located in chymotrypsinogen, and so *the oxyanion hole is incomplete in the zymogen.*

4. The conformational changes elsewhere in the molecule are very small. Thus, the switching on of enzymatic activity in a protein can be accomplished by discrete, highly localized conformational changes that are triggered by the hydrolysis of a single peptide bond.

THE GENERATION OF TRYPSIN FROM TRYPSINOGEN LEADS TO THE ACTIVATION OF OTHER ZYMOGENS

The structural changes accompanying the activation of trypsinogen, the precursor of the proteolytic enzyme trypsin, are somewhat different from those in the activation of chymotrypsinogen. X-ray analyses have shown that the conformation of four stretches of polypeptide, constituting about 15per cent of the molecule, changes markedly on activation. *These regions, called the activation domain, are very flexible in the zymogen, whereas they have a well-defined conformation in trypsin.* Furthermore, the oxyanion hole in trypsinogen is too far from histidine 57 to promote the formation of the tetrahedral transition state.

The digestion of proteins in the duodenum requires the concurrent action of several proteolytic enzymes, because each is specific for a limited number of side chains. Thus, the zymogens must be switched on at the same time. Coordinated control is achieved by the action of *trypsin as the common activator of all the pancreatic zymogens*—trypsinogen, chymotrypsinogen, proelastase, procarboxypeptidase, and prolipase, a lipid degrading enzyme. To produce active trypsin, the cells that line the duodenum secrete an enzyme, *enteropeptidase,* that hydrolyzes a unique lysine-isoleucine peptide bond in trypsinogen as the zymogen enters the duodenum from the pancreas. The small amount of trypsin produced in this way activates more trypsinogen and the other zymogens. Thus, *the formation of trypsin by enteropeptidase is the master activation step.*

SOME PROTEOLYTIC ENZYMES HAVE SPECIFIC INHIBITORS

The conversion of a zymogen into a protease by cleavage of a single peptide bond is a precise means of switching on enzymatic activity. However, this activation step is irreversible, and so a different mechanism is needed to stop proteolysis. Specific protease inhibitors accomplish this task. For example, *pancreatic trypsin inhibitor*, a 6-kd protein, inhibits trypsin by binding very tightly to its active site. The dissociation constant of the complex is 0.1 pM, which corresponds to a standard free energy of binding of about -18 kcal mol^{-1} (-75 kJ mol^{-1}). In contrast with nearly all known protein assemblies, this complex is not dissociated into its constituent chains by treatment with denaturing agents such as 8 M urea or 6 M guanidine hydrochloride.

The reason for the exceptional stability of the complex is that pancreatic trypsin inhibitor is a very effective substrate analog. X-ray analyses showed that the inhibitor lies in the active site of the enzyme, positioned such that the side chain of lysine 15 of this inhibitor interacts with the aspartate side chain in the specificity pocket of trypsin. In addition, there are many hydrogen bonds between the main chain of trypsin and that of its inhibitor. Furthermore, the carbonyl group of lysine 15 and the surrounding atoms of the inhibitor fit snugly in the active site of the enzyme. Comparison of the structure of the inhibitor bound to the enzyme with that of the free inhibitor reveals that *the structure is essentially unchanged on binding to the enzyme*. Thus, the inhibitor is preorganized into a structure that is highly complementary to the enzyme's active site. Indeed, the peptide bond between lysine 15 and alanine 16 in pancreatic trypsin inhibitor is cleaved but at a very slow rate: the half-life of the trypsin-inhibitor complex is several months. In essence, the inhibitor is a substrate, but its intrinsic structure is so nicely complementary to the enzyme's active site that it binds very tightly and is turned over slowly.

Why does trypsin inhibitor exist? Indeed, the amount of trypsin is much greater that that of the inhibitor. Under what circumstances is it beneficial to inhibit trypsin? Recall that trypsin activates other zymogens. Consequently, it is vital that even small amounts of trypsin be prevented from initiating the cascade prematurely. Trypsin molecules activated in the pancreas or pancreatic ducts could severely damage those tissues, leading to acute pancreatitis. Tissue necrosis may result from the activation of proteolytic enzymes (as well as prolipases) by trypsin, and hemorrhaging may result from its activation of elastase. We see here the physiological need for the tight binding of the inhibitor to trypsin.

Pancreatic trypsin inhibitor is not the only important protease inhibitor. *α_1-Antitrypsin* (also called *α_1-antiproteinase*), a 53-kd plasma protein, protects tissues from digestion by elastase, a secretory product of neutrophils (white blood cells that engulf bacteria). *Antielastase* would be a more accurate name for this inhibitor, because it blocks elastase much more effectively than it blocks trypsin. Like pancreatic trypsin inhibitor, α_1-antitrypsin blocks the action of target enzymes by binding nearly irreversibly to their active sites. Genetic disorders leading to a deficiency of α_1-antitrypsin show that this inhibitor is physiologically important. For example, the substitution of lysine for glutamate at residue 53 in the type Z mutant slows the secretion of this inhibitor from liver cells. Serum levels of the inhibitor are about 15per cent of normal in people homozygous for this defect. The consequence is that excess elastase destroys alveolar walls in the lungs by digesting elastic fibers and other connective-tissue proteins.

The resulting clinical condition is called *emphysema* (also known as *destructive lung disease*). People with emphysema must breathe much harder than normal people to exchange the same volume of air, because their alveoli are much less resilient than normal. Cigarette smoking markedly increases the likelihood that even a type Z heterozygote will develop emphysema. The reason is that smoke oxidizes methionine 358 of the inhibitor a residue essential for binding elastase. Indeed, this methionine side chain is the bait that selectively traps elastase. The *methionine sulfoxide* oxidation product, in contrast, does not lure elastase, a striking consequence of the insertion of just one oxygen atom into a protein. We will consider another protease inhibitor, antithrombin III, when we examine the control of blood clotting.

BLOOD CLOTTING IS ACCOMPLISHED BY A CASCADE OF ZYMOGEN ACTIVATIONS

Enzymatic cascades are often employed in biochemical systems to achieve a rapid response. In a cascade, an initial signal institutes a series of steps, each of which is catalyzed by an enzyme. At each step, the signal is amplified. For instance, if a signal molecule activates an enzyme that in turn activates 10 enzymes and each of the 10 enzymes in turn activates 10 additional enzymes, after four steps the original signal will have been amplified 10,000-fold. Blood clots are formed by a *cascade of zymogen activations:* the activated form of one clotting factor catalyzes the activation of the next. Thus, very small amounts of the initial factors suffice to trigger the cascade, ensuring a rapid response to trauma.

There are two means of initiating clotting: the intrinsic and extrinsic pathways. The *intrinsic clotting pathway* is activated by exposure of anionic surfaces on rupture of the endothelial lining of the blood vessels. These surfaces serve as binding sites for factors in the clotting cascade. Substances that are released from tissues as a consequence of trauma to them trigger the *extrinsic clotting pathway*. The extrinsic and intrinsic pathways converge on a common sequence of final steps to form a clot composed of the protein fibrin. The two pathways interact with each other in vivo. Indeed, both are needed for proper clotting, as evidenced by clotting disorders caused by a deficiency of a single protein in one of the pathways. Note that the active forms of the clotting factors are designated with a subscript "a."

FIBRINOGEN IS CONVERTED BY THROMBIN INTO A FIBRIN CLOT

The best-characterized part of the clotting process is the final step in the cascade: the conversion of fibrinogen into fibrin by thrombin, a proteolytic enzyme. Fibrinogen is made up of three globular units connected by two rods. This 340-kd protein consists of six chains: two each of Aα, Bβ, and γ. The rod regions are triple-stranded α-helical coiled coils, a recurring motif in proteins. Thrombin cleaves four *arginine-glycine peptide bonds* in the central globular region of fibrinogen. On cleavage, an A peptide of 18 residues from each of the two Aα chains and a B peptide of 20 residues from each of the two Bβ chains are released from the globular region. These A and B peptides are called *fibrinopeptides.* A fibrinogen molecule devoid of these fibrinopeptides is called a *fibrin monomer* and has the subunit structure $(\alpha\beta\gamma)_2$.

Fibrin monomers spontaneously assemble into ordered fibrous arrays called *fibrin.* Electron micrographs and low-angle x-ray patterns show that fibrin has a periodic structure that repeats every 23 nm. Higher-resolution images reveal the detailed structure of the fibrin monomer, how the fibrin monomers come together, and how this assembly is facilitated by removal of the fibrinopeptides. The homologous β and γ chains have globular domains at the carboxyl-terminal ends. These domains have binding "holes" that interact with peptides. The β domain is specific for sequences of the form H_3N^+-Gly-His-Arg-, whereas the γ domain binds H_3N^+-Gly-Pro-Arg-. Exactly these sequences (sometimes called "knobs") are exposed at the amino-terminal ends of the β and α chains, respectively, on thrombin cleavage. The knobs of the α subunits fit into the holes on the γ subunits of another monomer to form a protofibril. This protofibril is extended when the knobs of the β subunits fit into holes of β subunits of other protofibrils. Thus, analogous to the activation of chymotrypsinogen, peptide-bond cleavage exposes new amino termini that can participate in specific interactions. The newly formed clot is stabilized by the formation of amide bonds between the side chains of lysine and glutamine residues in different monomers.

This cross-linking reaction is catalyzed by *transglutaminase (factor XIII$_a$),* which itself is activated from the protransglutaminase form by thrombin.

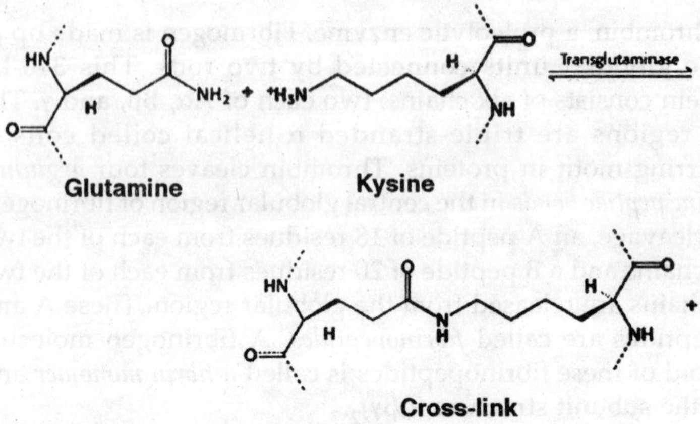

Transglutaminase

Glutamine Kysine

Cross-link

PROTHROMBIN IS READIED FOR ACTIVATION BY A VITAMIN K-DEPENDENT MODIFICATION

Thrombin is synthesized as a zymogen called *prothrombin,* which comprises four major domains, with the serine protease domain at its carboxyl terminus. The first domain is called a *gla domain,* whereas domains 2 and 3 are called *kringle domains.* These domains work in concert to keep prothrombin in an inactive form and to target it to appropriate sites for its activation by factor X_a (a serine protease) and factor V_a (a stimulatory protein). Activation is accomplished by proteolytic cleavage of the bond between arginine 274 and threonine 275 to release a fragment containing the first three domains and by cleavage of the bond between arginine 323 and isoleucine 324 (analogous to the key bond in chymotrypsinogen) to yield active thrombin.

Vitamin K has been known for many years to be essential for the synthesis of prothrombin and several other clotting factors. The results of studies of the abnormal prothrombin synthesized in the absence of vitamin K or in the presence of vitamin K antagonists, such as dicoumarol, revealed the mode of action of this vitamin.*Dicoumarol*is found in spoiled sweet clover and causes a fatal hemorrhagic disease in cattle fed on this hay. This coumarin derivative is used clinically as an *anticoagulant*to prevent thromboses in patients prone to clot formation. Dicoumarol and such related vitamin K antagonists as *warfarin*also serve as effective rat poisons. Cows fed dicoumarol synthesize an abnormal prothrombin that does not bind Ca^{2+}, in contrast with normal prothrombin. This difference was puzzling for some time because abnormal prothrombin has the same number of amino acid residues and gives the same amino acid analysis after acid hydrolysis as does normal prothrombin.

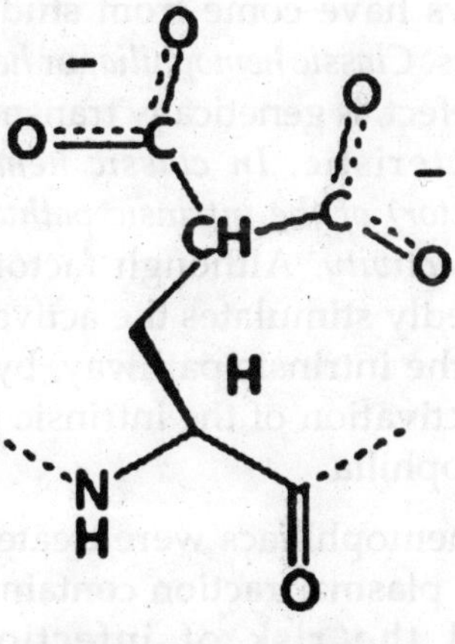

γ- Carboxyglutamate residue

Nuclear magnetic resonance studies revealed that normal prothrombin contains *γ-carboxyglutamate,* a previously unknown residue that evaded detection because its second carboxyl group is lost on acid hydrolysis. The abnormal prothrombin formed subsequent to the administration of anticoagulants lacks this modified amino acid. In fact, the first 10 glutamate residues in the amino-terminal region of prothrombin are carboxylated to γ-carboxyglutamate by a vitamin K-dependent enzyme system. *The vitamin K-dependent carboxylation reaction converts glutamate, a weak chelator of* Ca^{2+}, *into γ-carboxyglutamate, a much stronger chelator.* Prothrombin is thus able to bind Ca^{2+}, but what is the effect of this binding? The binding of Ca^{2+} by prothrombin anchors the zymogen to phospholipid membranes derived from blood platelets after injury. The binding of prothrombin to phospholipid surfaces is crucial because it brings prothrombin into close proximity to two clotting proteins that catalyze its conversion into thrombin. The proteolytic activation of prothrombin removes the calcium-binding domain and frees thrombin from the membrane so that it can cleave fibrinogen and other targets.

HEMOPHILIA REVEALED AN EARLY STEP IN CLOTTING

Some important breakthroughs in the elucidation of clotting path ways have come from studies of patients with bleeding disorders. *Classic hemophilia,* or *hemophilia A,* the best-known clotting defect, is genetically transmitted as a sex-linked recessive characteristic. *In classic hemophilia, factor VIII (antihemophilic factor) of the intrinsic pathway is missing or has markedly reduced activity.* Although factor VIII is not itself a protease, it markedly stimulates the activation of factor X, the final protease of the intrinsic pathway, by factor IX_a, a serine protease. Thus, activation of the intrinsic pathway is severely impaired in hemophilia.

In the past, hemophiliacs were treated with transfusions of a concentrated plasma fraction containing factor VIII. This therapy carried the risk of infection. Indeed, many hemophiliacs contracted hepatitis and AIDS. A safer preparation of factor VIII was urgently needed. With the use of biochemical purification and recombinant DNA techniques, the gene for factor VIII was isolated and expressed in cells grown in culture. Recombinant factor VIII purified from these cells has largely replaced plasma concentrates in treating hemophilia.

THE CLOTTING PROCESS MUST BE PRECISELY REGULATED

There is a fine line between hemorrhage and thrombosis. Clots must form rapidly yet remain confined to the area of injury. What are the mechanisms that normally limit clot formation to the site of injury? The lability of clotting factors contributes significantly to the control of clotting. Activated factors are short-lived because they are diluted by blood flow, removed by the liver, and degraded by proteases. For example, the stimulatory proteins factors V_a and VIII_a are digested by protein C, a protease that is switched on by the action of thrombin. *Thus, thrombin has a dual function: it catalyzes the formation of fibrin and it initiates the deactivation of the clotting cascade.*

Specific inhibitors of clotting factors are also critical in the termination of clotting. The most important one is *antithrombin III,* a plasma protein that inactivates thrombin by forming an irreversible complex with it. Antithrombin III resembles α_1-antitrypsin except that it inhibits thrombin much more strongly than it inhibits elastase. Antithrombin III also blocks other serine proteases in the clotting cascade—namely, factors XII_a, XI_a, IX_a, and X_a. The inhibitory action of antithrombin III is enhanced by *heparin,* a negatively charged polysaccharide found in mast cells near the walls of blood vessels and on the surfaces of endothelial cells. Heparin acts as an *anticoagulant* by increasing the rate of formation of irreversible complexes between antithrombin III and the serine protease clotting factors. Antitrypsin and antithrombin are *serpins,* a family of *ser*ine *p*rotease *in*hibitors.

The importance of the ratio of thrombin to antithrombin is illustrated in the case in which a 14-year-old boy died of a bleeding disorder because of a mutation in his α_1-antitrypsin, which normally inhibits elastase. Methionine 358 in α_1-antitrypsin's binding pocket for elastase was replaced by arginine, resulting in a change in specificity from an elastase inhibitor to a thrombin inhibitor. α_1-Antitrypsin activity normally increases markedly after injury to counteract excess elastase arising from stimulated neutrophils. The mutant α_1-antitrypsin caused the patient's thrombin activity to drop to such a low level that hemorrhage ensued. *We see here a striking example of how a change of a single residue in a protein can dramatically alter specificity and an example of the critical importance of having the right amount of a protease inhibitor.*

Antithrombin limits the extent of clot formation, but what happens to the clots themselves? Clots are not permanent structures but are designed to be dissolved when the structural integrity of damaged areas is restored. Fibrin is split by *plasmin,* a serine protease that hydrolyzes peptide bonds in the coiled-coil regions. Plasmin molecules can diffuse through aqueous channels in the porous fibrin clot to cut the accessible connector rods. Plasmin is formed by proteolytic activation of *plasminogen,* an inactive precursor that has a high affinity for the fibrin clots. This conversion is carried out by *tissue-type plasminogen activator* (TPA), a 72-kd protein that has a domain structure closely related to that of prothrombin.

However, a domain that targets TPA to fibrin clots replaces the membrane-targeting gla domain of prothrombin. The TPA bound to fibrin clots swiftly activates adhering plasminogen. In contrast, TPA activates free plasminogen very slowly. The gene for TPA has been cloned and expressed in cultured mammalian cells. The results of clinical studies have shown that TPA administered intravenously within an hour of the formation of a blood clot in a coronary artery markedly increases the likelihood of surviving a heart attack.

LIPIDS AND CELL MEMBRANES

The boundaries of cells are formed by *biological membranes,* the barriers that *define the inside and the outside of a cell.* These barriers prevent molecules generated inside the cell from leaking out and unwanted molecules from diffusing in; yet they also contain transport systems that allow specific molecules to be taken up and unwanted compounds to be removed from the cell. Such transport systems confer on membranes the important property of *selective permeability.*

Membranes are dynamic structures in which proteins float in a sea of lipids. The lipid components of the membrane form the permeability barrier, and protein components act as a transport system of pumps and channels that endow the membrane with selective permeability.

In addition to an external cell membrane (called the plasma membrane), eukaryotic cells also contain internal membranes that form the boundaries of organelles such as mitochondria, chloroplasts, peroxisomes, and lysosomes. Functional specialization in the course of evolution has been closely linked to the formation of such compartments. Specific systems have evolved to allow targeting of selected proteins into or through particular internal membranes and, hence, into specific organelles. External and internal membranes have essential features in common, and these essential features are the subject of this chapter.

Biological membranes serve several additional important functions indispensable for life, such as energy storage and information transduction, that are dictated by the proteins associated with them. In this chapter, we will examine the general properties of membrane proteins—how they can exist in the hydrophobic environment of the membrane while connecting two hydrophilic environments—and delay a discussion of the functions of these proteins to the next and later chapters.

MANY COMMON FEATURES UNDERLIE THE DIVERSITY OF BIOLOGICAL MEMBRANES

Membranes are as diverse in structure as they are in function. However, they do have in common a number of important attributes:

1. Membranes are *sheetlike structures,* only two molecules thick, that form *closed boundaries* between different compartments. The thickness of most membranes is between 60 Å (6 nm) and 100 Å (10 nm).
2. Membranes consist mainly of *lipids* and *proteins*. Their mass ratio ranges from 1:4 to 4:1. Membranes also contain *carbohydrates* that are linked to lipids and proteins.
3. Membrane lipids are relatively small molecules that have both *hydrophilic* and *hydrophobic* moieties. These lipids spontaneously form *closed bimolecular sheets* in aqueous media. These *lipid bilayers* are barriers to the flow of polar molecules.

4. *Specific proteins mediate distinctive functions of membranes.* Proteins serve as pumps, channels, receptors, energy transducers, and enzymes. Membrane proteins are embedded in lipid bilayers, which create suitable environments for their action.
5. Membranes are *noncovalent assemblies.* The constituent protein and lipid molecules are held together by many noncovalent interactions, which are cooperative.
6. Membranes are *asymmetric.* The two faces of biological membranes always differ from each other.
7. Membranes are *fluid structures.* Lipid molecules diffuse rapidly in the plane of the membrane, as do proteins, unless they are anchored by specific interactions. In contrast, lipid molecules and proteins do not readily rotate across the membrane. Membranes can be regarded as *two-dimensional solutions of oriented proteins and lipids.*
8. Most cell membranes are *electrically polarized,* such that the inside is negative [typically - 60 millivolts (mV)]. Membrane potential plays a key role in transport, energy conversion, and excitability.

FATTY ACIDS ARE KEY CONSTITUENTS OF LIPIDS

Among the most biologically significant properties of lipids are their hydrophobic properties. These properties are mainly due to a particular component of lipids: fatty acids, or simply fats. Fatty acids also play important roles in signal-transduction pathways.

THE NAMING OF FATTY ACIDS

Fatty acids are hydrocarbon chains of various lengths and degrees of unsaturation that terminate with carboxylic acid groups. The systematic name for a fatty acid is derived from the name of its parent hydrocarbon by the substitution of *oic* for the final *e*. For example, the C_{18} saturated fatty acid is called *octadecanoic acid* because the parent hydrocarbon is octadecane. A C_{18} fatty acid with one double bond is called octadec*enoic* acid; with two double bonds, octadeca*dienoic* acid; and with three double bonds, octadeca*trienoic* acid. The notation 18:0 denotes a C_{18} fatty acid with no double bonds, whereas 18:2 signifies that there are two double bonds. The structures of the ionized forms of two common fatty acids—palmitic acid (C_{16}, saturated) and oleic acid (C_{18}, monounsaturated).

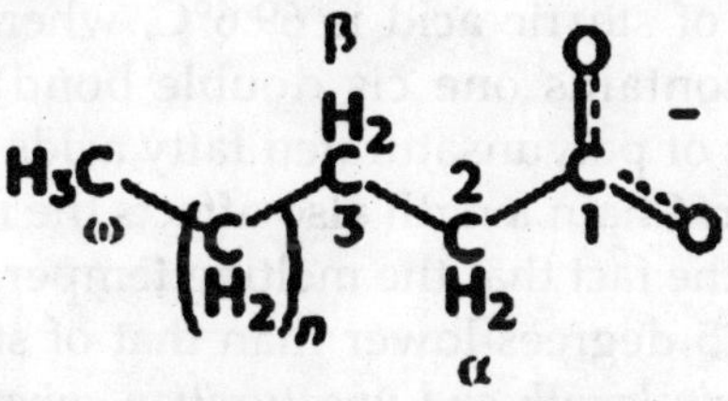

Fatty acid carbon atoms are numbered starting at the carboxyl terminus, as shown in the margin. Carbon atoms 2 and 3 are often referred to as α and β, respectively. The methyl carbon atom at the distal end of the chain is called the *ω-carbon atom.* The position of a double bond is represented by the symbol Δ followed by a superscript number. For example, *cis*-Δ^9 means that there is a cis double bond between carbon atoms 9 and 10; *trans*-Δ^2 means that there is a trans double bond between carbon atoms 2 and 3. Alternatively, the position of a double bond can be denoted by counting from the distal end, with the ω-carbon atom (the methyl carbon) as number 1. An ω-3 fatty acid, for example, has the structure shown in the margin. Fatty acids are ionized at physiological pH, and so it is appropriate to refer to them according to their carboxylate form: for example, palmitate or hexadecanoate.

FATTY ACIDS VARY IN CHAIN LENGTH AND DEGREE OF UNSATURATION

Fatty acids in biological systems usually contain an even number of carbon atoms, typically between 14 and 24. The 16- and 18-carbon fatty acids are most common. The hydrocarbon chain is almost invariably unbranched in animal fatty acids.

The alkyl chain may be saturated or it may contain one or more double bonds. The configuration of the double bonds in most unsaturated fatty acids is cis. The double bonds in polyunsaturated fatty acids are separated by at least one methylene group.

The properties of fatty acids and of lipids derived from them are markedly dependent on chain length and degree of saturation. Unsaturated fatty acids have lower melting points than saturated fatty acids of the same length. For example, the melting point of stearic acid is 69.6°C, whereas that of oleic acid (which contains one cis double bond) is 13.4°C. The melting points of polyunsaturated fatty acids of the C_{18} series are even lower. Chain length also affects the melting point, as illustrated by the fact that the melting temperature of palmitic acid (C_{16}) is 6.5 degrees lower than that of stearic acid (C_{18}). Thus, *short chain length and unsaturation enhance the fluidity of fatty acids and of their derivatives.*

THREE COMMON TYPES OF MEMBRANE LIPIDS

Lipids differ markedly from the other groups of biomolecules considered thus far. By definition, *lipids are water-insoluble biomolecules that are highly soluble in organic solvents such as chloroform.* Lipids have a variety of biological roles: they serve as fuel molecules, highly concentrated energy stores, signal molecules, and components of membranes. The first three roles of lipids will be discussed in later chapters. Here, our focus is on lipids as membrane constituents. The three major kinds of membrane lipids are *phospho-lipids, glycolipids,* and *cholesterol.* We begin with lipids found in eukaryotes and bacteria. The lipids in archaea are distinct, although they have many features related to their membrane-forming function in common with lipids of other organisms.

PHOSPHOLIPIDS ARE THE MAJOR CLASS OF MEMBRANE LIPIDS

Phospholipids are abundant in all biological membranes. A phospholipid molecule is constructed from four components: fatty acids, a platform to which the fatty acids are attached, a phosphate, and an alcohol attached to the phosphate. The fatty acid components provide a hydrophobic barrier, whereas the remainder of the molecule has hydrophilic properties to enable interaction with the environment.

The platform on which phospholipids are built may be *glycerol,* a 3- carbon alcohol, or *sphingosine,* a more complex alcohol. Phospholipids derived from glycerol are called *phosphoglycerides.* A phosphoglyceride consists of a glycerol backbone to which two fatty acid chains and a phosphorylated alcohol are attached.

In phosphoglycerides, the hydroxyl groups at C-1 and C-2 of glycerol are esterified to the carboxyl groups of the two fatty acid chains. The C-3 hydroxyl group of the glycerol backbone is esterified to phosphoric acid. When no further additions are made, the resulting compound is *phosphati-date (diacylglycerol 3-phosphate),* the simplest phosphoglyceride. Only small amounts of phosphatidate are present in membranes. However, the molecule is a key intermediate in the biosynthesis of the other phosphoglycerides.

The major phosphoglycerides are derived from phosphatidate by the formation of an ester bond between the phosphate group of phosphatidate and the hydroxyl group of one of several alcohols. The common alcohol moieties of phosphoglycerides are the amino acid serine, ethanolamine, choline, glycerol, and the inositol.

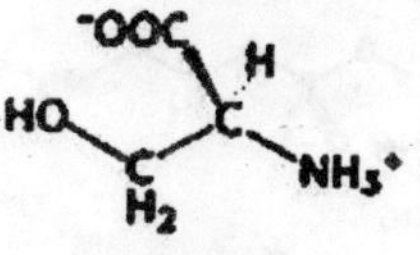

Serine 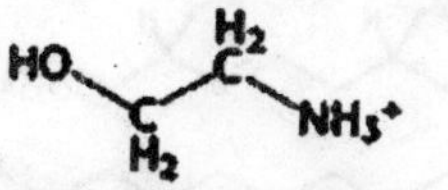 Ethanolamine 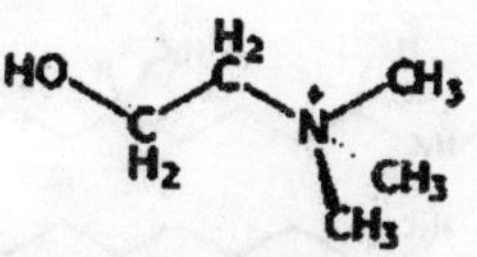 Choline

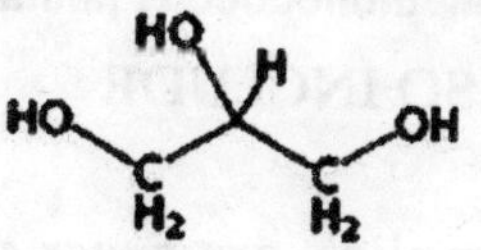

Glycerol 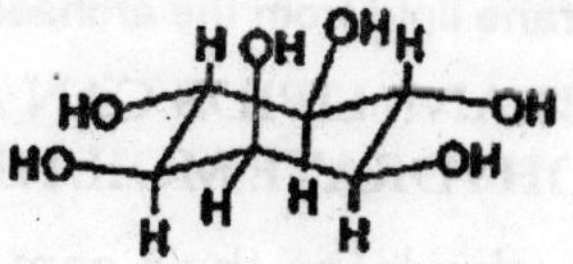 Inositol

The structural formulas of phosphatidyl choline and the other principal phosphoglycerides—namely, phosphatidyl ethanolamine, phosphatidyl serine, phosphatidyl inositol, and diphosphatidyl glycerol.

Sphingomyelin is a phospholipid found in membranes that is not derived from glycerol. Instead, the backbone in sphingomyelin is *sphingosine*, an amino alcohol that contains a long, unsaturated hydrocarbon chain.. In sphingomyelin, the amino group of the sphingosine backbone is linked to a fatty acid by an amide bond. In addition, the primary hydroxyl group of sphingosine is esterified to phosphoryl choline.

Archaeal Membranes are Built from Ether Lipids with Branched Chains

The membranes of archaea differ in composition from those of eukaryotes or bacteria in three important ways. Two of these differences clearly relate to the hostile living conditions of many archaea. First, the nonpolar chains are joined to a glycerol backbone by *ether* rather than ester linkages. The ether linkage is more resistant to hydrolysis. Second, the alkyl chains are *branched* rather than linear. They are built up from repeats of a fully saturated five-carbon fragment. These branched, saturated hydrocarbons are more resistant to oxidation. The ability of archaeal lipids to resist hydrolysis and oxidation may help these organisms to withstand the extreme conditions, such as high temperature, low pH, or high salt concentration, under which some of these archaea grow. Finally, the stereochemistry of the central glycerol is inverted compared with that shown.

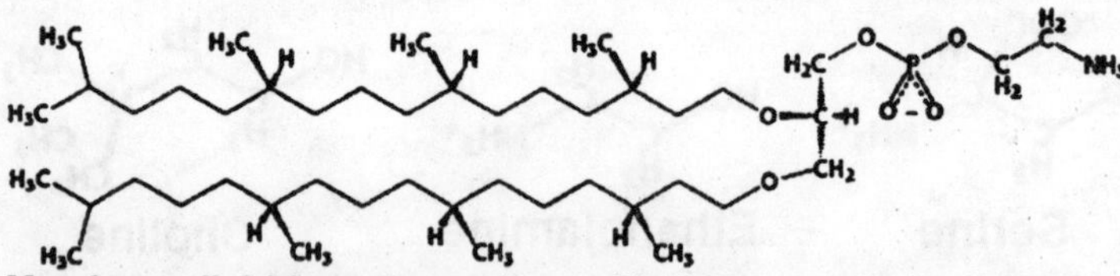

Membrane lipid from the archaeon Methonococcus jannaschff

MEMBRANE LIPIDS CAN ALSO INCLUDE CARBOHYDRATE MOIETIES

Glycolipids, as their name implies, are *sugar-containing lipids*. Like sphingomyelin, the glycolipids in animal cells are derived from sphingosine. The amino group of the sphingosine backbone is acylated by a fatty acid, as in sphingomyelin. Glycolipids differ from sphingomyelin in the identity of the unit that is linked to the primary hydroxyl group of the sphingosine backbone. In glycolipids, one or more sugars (rather than phosphoryl choline) are attached to this group. The simplest glycolipid, called a *cerebroside,* contains a single sugar residue, either glucose or galactose.

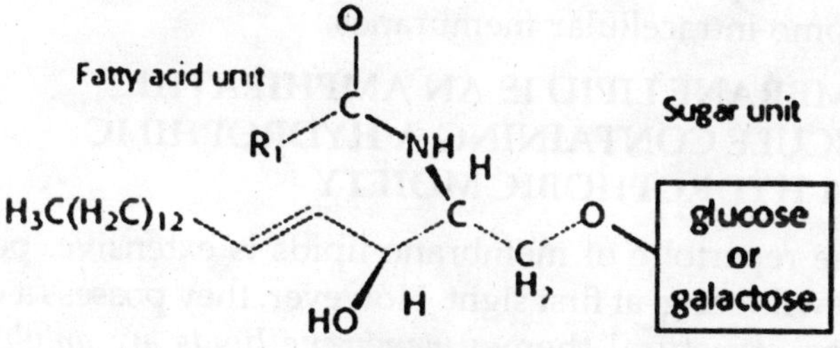

Cerebroside (a glycolipid)

More complex glycolipids, such as *gangliosides,* contain a branched chain of as many as seven sugar residues. Glycolipids are oriented in a completely asymmetric fashion with the *sugar residues always on the extracellular side of the membrane.*

CHOLESTEROL IS A LIPID BASED ON A STEROID NUCLEUS

Cholesterol is a lipid with a structure quite different from that of phospholipids. It is a steroid, built from four linked hydrocarbon rings.

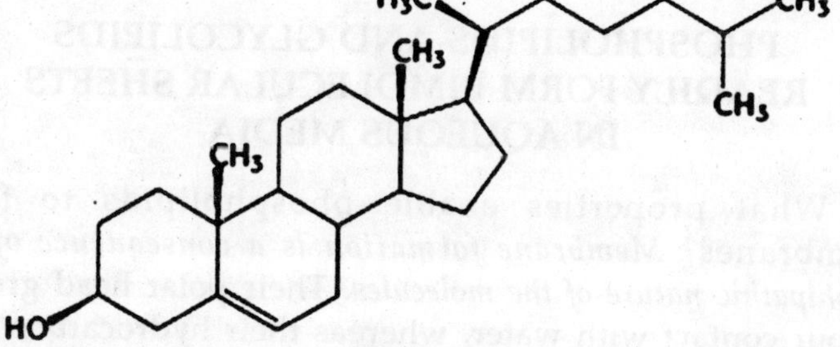

Cholesterol

A hydrocarbon tail is linked to the steroid at one end, and a hydroxyl group is attached at the other end. In membranes, the molecule is oriented parallel to the fatty acid chains of the phospholipids, and the hydroxyl group interacts with the nearby phospholipid head groups. Cholesterol is absent from prokaryotes but is found to varying degrees in virtually all animal membranes. It constitutes almost 25per cent of the membrane lipids in certain nerve cells but is essentially absent from some intracellular membranes.

A MEMBRANE LIPID IS AN AMPHIPATHIC MOLECULE CONTAINING A HYDROPHILIC AND A HYDROPHOBIC MOIETY

The repertoire of membrane lipids is extensive, perhaps even bewildering, at first sight. However, they possess a critical common structural theme: *membrane lipids are amphipathic molecules* (amphiphilic molecules). A membrane lipid contains both a *hydrophilic* and a *hydrophobic* moiety.

Let us look at a model of a phosphoglyceride, such as phosphatidyl choline. Its overall shape is roughly rectangular. The two hydrophobic fatty acid chains are approximately parallel to each other, whereas the hydrophilic phosphoryl choline moiety points in the opposite direction. Sphingomyelin has a similar conformation, as does the archaeal lipid depicted. Therefore, the following shorthand has been adopted to represent these membrane lipids: the hydrophilic unit, also called the *polar head group*, is represented by a circle, whereas the hydrocarbon tails are depicted by straight or wavy lines.

PHOSPHOLIPIDS AND GLYCOLIPIDS READILY FORM BIMOLECULAR SHEETS IN AQUEOUS MEDIA

What properties enable phospholipids to form membranes? *Membrane formation is a consequence of the amphipathic nature of the molecules.* Their polar head groups favour contact with water, whereas their hydrocarbon tails interact with one another, in preference to water. How can molecules with these preferences arrange themselves in aqueous solutions? One way is to form a *micelle,* a globular structure in which polar head groups are surrounded by water and hydrocarbon tails are sequestered inside, interacting with one another.

Alternatively, the strongly opposed preferences of the hydrophilic and hydrophobic moieties of membrane lipids can be satisfied by forming a *lipid bilayer,* composed of two lipid sheets. A lipid bilayer is also called a *bimolecular sheet.* The hydrophobic tails of each individual sheet interact with one another, forming a hydrophobic interior that acts as a permeability barrier. The hydrophilic head groups interact with the aqueous medium on each side of the bilayer. The two opposing sheets are called leaflets.

The favoured structure for most phospholipids and glycolipids in aqueous media is a bimolecular sheet rather than a micelle. The reason is that the two fatty acyl chains of a phospholipid or a glycolipid are too bulky to fit into the interior of a micelle. In contrast, salts of fatty acids (such as sodium palmitate, a constituent of soap), which contain only one chain, readily form micelles. *The formation of bilayers instead of micelles by phospholipids is of critical biological importance.* A micelle is a limited structure, usually less than 20 nm (200 Å) in diameter. In contrast, a bimolecular sheet can have macroscopic dimensions, such as a millimeter (10^6 nm, or 10^7 Å). Phospholipids and related molecules are important membrane constituents because they readily form extensive bimolecular sheets.

The formation of lipid bilayers is a *self-assembly process.* In other words, the structure of a bimolecular sheet is inherent in the structure of the constituent lipid molecules. The growth of lipid bilayers from phospholipids is a rapid and spontaneous process in water. *Hydrophobic interactions are the major driving force for the formation of lipid bilayers.* Recall that hydrophobic interactions also play a dominant role in the folding of proteins and in the stacking of bases in nucleic acids. Water molecules are released from the hydrocarbon tails of membrane lipids as these tails become sequestered in the nonpolar interior of the bilayer. Furthermore, *van der Waals attractive forces between the hydrocarbon tails favour close packing of the tails.* Finally, there are *electrostatic and hydrogen-bonding attractions between the polar head groups and water molecules.* Thus, lipid bilayers are stabilized by the full array of forces that mediate molecular interactions in biological systems.

Because lipid bilayers are held together by many *reinforcing, noncovalent interactions (predominantly hydrophobic),* they are *cooperative structures.* These hydrophobic interactions have three significant biological consequences:

1. Lipid bilayers have an inherent tendency to be *extensive;*
2. Lipid bilayers will tend to *close on themselves* so that there are no edges with exposed hydrocarbon chains, and so they form compartments; and
3. Lipid bilayers are *self-sealing* because a hole in a bilayer is energetically unfavourable.

LIPID VESICLES CAN BE FORMED FROM PHOSPHOLIPIDS

The propensity of phospholipids to form membranes has been used to create an important experimental and clinical tool. *Lipid vesicles,* or *liposomes,* aqueous compartments enclosed by a lipid bilayer can be used to study membrane permeability or to deliver chemicals to cells. Liposomes are formed by suspending a suitable lipid, such as phosphatidyl choline, in an aqueous medium, and then *sonicating* (i.e., agitating by high-frequency sound waves) to give a dispersion of closed vesicles that are quite uniform in size. Vesicles formed by these methods are nearly spherical in shape and have a diameter of about 50 nm (500 Å). Larger vesicles (of the order of 1 μm, or 10^4 Å, in diameter) can be prepared by slowly evapourating the organic solvent from a suspension of phospholipid in a mixed solvent system.

Ions or molecules can be trapped in the aqueous compartments of lipid vesicles by forming the vesicles in the presence of these substances. For example, 50-nm-diameter vesicles formed in a 0.1 M glycine solution will trap about 2000 molecules of glycine in each inner aqueous compartment. These glycine-containing vesicles can be separated from the surrounding solution of glycine by dialysis or by gel-filtration chromatography. The permeability of the bilayer membrane to glycine can then be determined by measuring the rate of efflux of glycine from the inner compartment of the vesicle to the ambient solution. Specific membrane proteins can be solubilized in the presence of detergents and then added to the phospholipids from which liposomes will be formed. Protein-liposome complexes provide valuable experimental tools for examining a range of membrane-protein functions.

Experiments are underway to develop clinical uses for liposomes. For example, liposomes containing drugs or DNA for gene-therapy experiments can be injected into patients. These liposomes fuse with the plasma membrane of many kinds of cells, introducing into the cells the molecules that they contain. Drug delivery with liposomes also alters the distribution of a drug within the body and often lessens its toxicity. For instance, less of the drug is distributed to normal tissues because longcirculating liposomes have been shown to concentrate in regions of increased blood circulation, such as solid tumors and sites of inflammation. Moreover, the selective fusion of lipid vesicles with particular kinds of cells is a promising means of controlling the delivery of drugs to target cells.

Another well-defined synthetic membrane is a *planar bilayer membrane.* This structure can be formed across a 1-mm hole in a partition between two aqueous compartments by dipping a fine paintbrush into a membrane-forming solution, such as phosphatidyl choline in decane. The tip of the brush is then stroked across a hole (1 mm in diameter) in a partition between two aqueous media. The lipid film across the hole thins spontaneously into a lipid bilayer. The electrical conduction properties of this macroscopic bilayer membrane are readily studied by inserting electrodes into each aqueous compartment. For example, its permeability to ions is determined by measuring the current across the membrane as a function of the applied voltage.

LIPID BILAYERS ARE HIGHLY IMPERMEABLE TO IONS AND MOST POLAR MOLECULES

The results of permeability studies of lipid vesicles and electrical-conductance measurements of planar bilayers have shown that *lipid bilayer membranes have a very low permeability for ions and most polar molecules.* Water is a conspicuous exception to this generalization; it readily traverses such membranes because of its small size, high concentration, and lack of a complete charge. The range of measured permeability coefficients is very wide. For example, Na^+ and K^+ traverse these membranes 10^9 times as slowly as does H_2O. Tryptophan, a zwitterion at pH 7, crosses the membrane 10^3 times as slowly as does indole, a structurally related molecule that lacks ionic groups. In fact, *the permeability coefficients of small molecules are correlated with their solubility in a nonpolar solvent relative to their solubility in water.* This relation suggests that a small molecule might traverse a lipid bilayer membrane in the following way: first, it sheds its solvation shell of water; then, it becomes dissolved in the hydrocarbon core of the membrane; finally, it diffuses through this core to the other side of the membrane, where it becomes resolvated by water. An ion such as Na^+ tra-verses membranes very slowly because the replacement of its coordination shell of polar water molecules by nonpolar interactions with the membrane interior is highly unfavourable energetically.

PROTEINS CARRY OUT MOST MEMBRANE PROCESSES

We now turn to membrane proteins, which are responsible for most of the dynamic processes carried out by membranes. Membrane lipids form a permeability barrier and thereby establish compartments, whereas *specific proteins mediate nearly all other membrane functions.* In particular, proteins transport chemicals and information across a membrane. Membrane lipids create the appropriate environment for the action of such proteins.

Membranes differ in their protein content. Myelin, a membrane that serves as an insulator around certain nerve fibers, has a low content of protein (18per cent). Relatively pure lipids are well suited for insulation. In contrast, the plasma membranes or exterior membranes of most other cells are much more active. They contain many pumps, channels, receptors, and enzymes. The protein content of these plasma membranes is typically 50per cent. Energy-transduction membranes, such as the internal membranes of mitochondria and chloroplasts, have the highest content of protein, typically 75per cent.

The protein components of a membrane can be readily visualized by *SDS-polyacrylamide gel electrophoresis*. As discussed earlier the electrophoretic mobility of many proteins in SDS-containing gels depends on the mass rather than on the net charge of the protein. The gel-electrophoresis patterns of three membranes—the plasma membrane of erythrocytes, the photoreceptor membrane of retinal rod cells, and the sarcoplasmic reticulum membrane of muscle—are shown in figure. It is evident that each of these three membranes contains many proteins but has a distinct protein composition. In general, *membranes performing different functions contain different repertoires of proteins.*

PROTEINS ASSOCIATE WITH THE LIPID BILAYER IN A VARIETY OF WAYS

The ease with which a protein can be dissociated from a membrane indicates how intimately it is associated with the membrane. Some membrane proteins can be solubilized by relatively mild means, such as extraction by a solution of high ionic strength (e.g., 1 M NaCl). Other membrane proteins are bound much more tenaciously; they can be solubilized only by using a detergent or an organic solvent. Membrane proteins can be classified as being either *peripheral* or *integral* on the basis of this difference in dissociability. *Integral membrane proteins* interact extensively with the hydrocarbon chains of membrane lipids, and so only agents that compete for these nonpolar interactions can release them. In fact, most integral membrane proteins span the lipid bilayer. In contrast, *peripheral membrane proteins* are bound to membranes primarily by electrostatic and hydrogen-bond interactions with the head groups of lipids. These polar interactions can be disrupted by adding salts or by changing the pH. Many peripheral membrane proteins are bound to the surfaces of integral proteins, on either the cytosolic or the extracellular side of the membrane. Others are anchored to the lipid bilayer by a covalently attached hydrophobic chain, such as a fatty acid.

PROTEINS INTERACT WITH MEMBRANES IN A VARIETY OF WAYS

Although membrane proteins are more difficult to purify and crystallize than are water-soluble proteins, researchers using x-ray crystallographic or electron microscopic methods have determined the three-dimensional structures of more than 20 such proteins at sufficiently high resolution to discern the molecular details. As noted the structures of membrane proteins differ from those of soluble proteins with regard to the distribution of hydrophobic and hydrophilic groups. We will consider the structures of three membrane proteins in some detail.

PROTEINS CAN SPAN THE MEMBRANE WITH ALPHA HELICES

The first membrane protein that we consider is the archaeal protein *bacteriorhodopsin,* shown in figure. This protein acts in energy transduction by using light energy to transport protons from inside the cell to outside. The proton gradient generated in this way is used to form ATP. Bacteriorhodopsin is built almost entirely of α helices; seven closely packed α helices, arranged almost perpendicularly to the plane of the membrane, span its 45-Å width. Examination of the primary structure of bacteriorhodopsin reveals that most of the amino acids in these membrane-spanning α helices are nonpolar and only a very few are charged. This distribution of nonpolar amino acids is sensible because these residues are either in contact with the hydrocarbon core of the membrane or with one another. *Membrane-spanning α helices are the most common structural motif in membrane proteins.* Such regions can often be detected from amino acid sequence information alone.

A CHANNEL PROTEIN CAN BE FORMED FROM BETA STRANDS

Porin, a protein from the outer membranes of bacteria such as *E. coli* and *Rhodo-bacter capsulatus*, represents a class of membrane proteins with a completely different type of structure. Structures of this type are built from β strands and contain essentially no α helices.

The arrangement of β strands is quite simple: each strand is hydrogen bonded to its neighbour in an antiparallel arrangement, forming a single *β* sheet. The β sheet curls up to form a hollow cylinder that functions as the active unit. As its name suggests, porin forms pores, or channels, in the membranes. A pore runs through the centre of each cylinder-like protein. The outside surface of porin is appropriately nonpolar, given that it interacts with the hydrocarbon core of the membrane. In contrast, the inside of the channel is quite hydrophilic and is filled with water. This arrangement of nonpolar and polar surfaces is accomplished by the alternation of hydrophobic and hydrophilic amino acids along each β strand.

EMBEDDING PART OF A PROTEIN IN A MEMBRANE CAN LINK IT TO THE MEMBRANE SURFACE

The structure of the membrane-bound enzyme prostaglandin H_2 synthase-1 reveals a rather different role for α helices in protein-membrane associations. This enzyme catalyzes the conversion of arachidonic acid into prostaglandin H_2 in two steps: a cyclooxygenase reaction and a peroxidase reaction. Prostaglandin H_2 promotes inflammation and modulates gastric acid secretion. The enzyme that produces prostaglandin H_2 is a homodimer with a rather complicated structure consisting primarily of α helices. Unlike bacteriorhodopsin, this protein is not largely embedded in the membrane but, instead, lies along the outer surface of the membrane firmly bound by a set of α helices with hydrophobic surfaces that extend from the bottom of the protein into the membrane. This linkage is sufficiently strong that only the action of detergents can release the protein from the membrane. Thus, this enzyme is classified as an integral membrane protein, although it is not a membrane-spanning protein.

The localization of prostaglandin H_2 synthase-l in the membrane is crucial to its function. The substrate for this enzyme, arachidonic acid, is a hydrophobic molecule generated by the hydrolysis of membrane lipids. Arachidonic acid reaches the active site of the enzyme from the membrane without entering an aqueous environment by traveling through a hydrophobic channel in the protein. Indeed, nearly all of us have experienced the importance of this channel: drugs such as aspirin and ibuprofen block the channel and prevent prostaglandin synthesis by inhibiting the cyclooxygenase activity of the synthase. In the case of aspirin, the drug acts through the transfer of an acetyl group from the aspirin to a serine residue (Ser 530) that lies along the path to the active site.

Two important features emerge from our examination of these three examples of membrane protein structure. First, the parts of the protein that interact with the hydrophobic parts of the membrane are coated with nonpolar amino acid side chains, whereas those parts that interact with the aqueous environment are much more hydrophilic. Second, the structures positioned within the membrane are quite regular and, in particular, all backbone hydrogen-bond donors and acceptors participate in hydrogen bonds. *Breaking a hydrogen bond within a membrane is quite unfavourable, because little or no water is present to compete for the polar groups.*

SOME PROTEINS ASSOCIATE WITH MEMBRANES THROUGH COVALENTLY ATTACHED HYDROPHOBIC GROUPS

The membrane proteins considered thus far associate with the membrane through surfaces generated by hydrophobic

amino acid side chains. However, even otherwise soluble proteins can associate with membranes if the association is mediated by hydrophobic groups attached to the proteins. Three such groups are shown in figure:

1. A palmitoyl group attached to a specific cysteine residue by a thioester bond,
2. A farnesyl group attached to a cysteine residue at the carboxyl terminus, and
3. A glycolipid structure termed a glycosyl phosphatidyl inositol (GPI) anchor attached to the carboxyl terminus. These modifications are attached by enzyme systems that recognize specific signal sequences near the site of attachment.

TRANSMEMBRANE HELICES CAN BE ACCURATELY PREDICTED FROM AMINO ACID SEQUENCES

Many membrane proteins, like bacteriorhodopsin, employ α helices to span the hydrophobic part of a membrane. As noted earlier, typically most of the residues in these α helices are nonpolar and almost none of them are charged. Can we use this information to identify putative membrane-spanning regions from sequence data alone? One approach to identifying transmembrane helices is to ask whether a postulated helical segment is likely to be most stable in a hydrocarbon milieu or in water. Specifically, we want to estimate the free-energy change when a helical segment is transferred from the interior of a membrane to water. For example, the transfer of a poly-l-arginine helix, a homopolymer of a positively charged amino acid, from the interior of a membrane to water would be highly favourable [-12.3 kcal mol^{-1} (-51.5 kJ mol^{-1}) per arginine residue in the helix], whereas the transfer of a poly-l-phenylalanine helix, a homopolymer of a hydrophobic amino acid, would be unfavourable [+3.7 kcal mol^{-1} (+15.5 kJ mol^{-1}) per phenylalanine residue in the helix].

The hydrocarbon core of a membrane is typically 30 Å wide, a length that can be traversed by an α helix consisting of 20 residues. We can take the amino acid sequence of a protein and estimate the free-energy change that takes place when a hypothetical α helix formed of residues 1 through 20 is transferred from the membrane interior to water. The same calculation can be made for residues 2 through 21, 3 through 22, and so forth, until we reach the end of the sequence. The span of 20 residues chosen for this calculation is called a *window*. The free-energy change for each window is plotted against the first amino acid at the window to create a *hydropathy plot*. Empirically, a peak of +20 kcal mol^{-1} (+84 kJ mol^{-1}) or more in a hydropathy plot based on a window of 20 residues indicates that a polypeptide segment could be a membrane-spanning α helix. For example, glycophorin, a protein found in the membranes of red blood cells, is predicted by this criterion to have one membrane-spanning helix, in agreement with experimental findings. It should be noted, however, that a peak in the hydropathy plot does not prove that a segment is a transmembrane helix. Even soluble proteins may have highly nonpolar regions. Conversely, some membrane proteins contain transmembrane structural features (such as membrane-spanning β strands) that escape detection by these plots.

LIPIDS AND MANY MEMBRANE PROTEINS DIFFUSE RAPIDLY IN THE PLANE OF THE MEMBRANE

Biological membranes are not rigid, static structures. On the contrary, lipids and many membrane proteins are constantly in lateral motion, a process called *lateral diffusion*. The rapid lateral movement of membrane proteins has been visualized by means of fluorescence microscopy through the use of the technique of *fluorescence recovery after photobleaching*. First, a cell-surface component is specifically labeled with a fluorescent chromophore. A small region of the cell surface (~3 μm^2) is viewed through a fluorescence microscope. The fluorescent molecules in this region are then destroyed (bleached) by a very intense light pulse from a laser. The fluorescence of this region is subsequently monitored as a function of time by using a light level sufficiently low to prevent further bleaching. If the labeled component is mobile, bleached molecules leave and unbleached molecules enter the illuminated region, which results in an increase in the fluorescence intensity. The rate of recovery of fluorescence depends on the lateral mobility of the fluorescence-labeled component, which can be expressed in terms of a diffusion coefficient, *D*. The average distance *s* traversed in time *t* depends on *D* according to the expression

The diffusion coefficient of lipids in a variety of membranes is about 1 $\mu m^2\ s^{-1}$. Thus, a phospholipid molecule diffuses an average distance of 2 μm in 1 s. This rate means that *a lipid molecule can travel from one end of a bacterium to the other in a second.* The magnitude of the observed diffusion coefficient indicates that the viscosity of the membrane is about 100 times that of water, rather like that of olive oil.

In contrast, proteins vary markedly in their lateral mobility. *Some proteins are nearly as mobile as lipids, whereas others are virtually immobile.* For example, the photoreceptor protein rhodopsin a very mobile protein, has a diffusion coefficient of 0.4 $\mu m^2\ s^{-1}$. The rapid movement of rhodopsin is essential for fast signaling. At the other extreme is fibronectin, a peripheral glycoprotein that interacts with the extracellular matrix. For fibronectin, *D* is less than $10^{-4}\ \mu m^2\ s^{-1}$. Fibronectin has a very low mobility because it is anchored to actin filaments on the inside of the plasma membrane through *integrin,* a transmembrane protein that links the extracellular matrix to the cytoskeleton.

THE FLUID MOSAIC MODEL ALLOWS LATERAL MOVEMENT BUT NOT ROTATION THROUGH THE MEMBRANE

On the basis of the dynamic properties of proteins in membranes, S. Jonathan Singer and Garth Nicolson proposed the concept of a *fluid mosaic model* for the overall organization of biological membranes in 1972. The essence of their model is that *membranes are two-dimensional solutions of oriented lipids and globular proteins.* The lipid bilayer has a dual role: it is both a *solvent* for integral membrane proteins and a *permeability barrier.* Membrane proteins are free to diffuse laterally in the lipid matrix unless restricted by special interactions.

Although the lateral diffusion of membrane components can be rapid, the spontaneous rotation of lipids from one face of a membrane to the other is a very slow process. The transition of a molecule from one membrane surface to the other is called *transverse diffusion* or *flip-flop*. The flip-flop of phospholipid molecules in phosphatidyl choline vesicles has been directly measured by electron spin resonance techniques, which show that *a phospholipid molecule flip-flops once in several hours.* Thus, a phospholipid molecule takes about 10^9 times as long to flip-flop across a membrane as it takes to diffuse a distance of 50 Å in the lateral direction. The free-energy barriers to flip-flopping are even larger for protein molecules than for lipids because proteins have more extensive polar regions. In fact, the flip-flop of a protein molecule has not been observed. Hence, *membrane asymmetry can be preserved for long periods.*

MEMBRANE FLUIDITY IS CONTROLLED BY FATTY ACID COMPOSITION AND CHOLESTEROL CONTENT

Many membrane processes, such as transport or signal transduction, depend on the fluidity of the membrane lipids, which in turn depends on the properties of fatty acid chains, which can exist in an ordered, rigid state or in a relatively disordered, fluid state. The transition from the rigid to the fluid state occurs rather abruptly as the temperature is raised above T_m, the melting temperature. *This transition temperature depends on the length of the fatty acyl chains and on their degree of unsaturation.* The presence of saturated fatty acyl residues favours the rigid state because their straight hydrocarbon chains interact very favourably with each other. On the other hand, *a cis double bond produces a bend in the hydrocarbon chain. This bend interferes with a highly ordered packing of fatty acyl chains, and so* T_m *is lowered.* The length of the fatty acyl chain also affects the transition temperature. Long hydrocarbon chains interact more strongly than do short ones. Specifically, each additional

$-CH_2-$ group makes a favourable contribution of about -0.5 kcal mol^{-1} (-2.1 kJ mol^{-1}) to the free energy of interaction of two adjacent hydrocarbon chains.

Bacteria regulate the fluidity of their membranes by varying the number of double bonds and the length of their fatty acyl chains. For example, the ratio of saturated to unsaturated fatty acyl chains in the *E. coli* membrane decreases from 1.6 to 1.0 as the growth temperature is lowered from 42°C to 27°C. This decrease in the proportion of saturated residues prevents the membrane from becoming too rigid at the lower temperature.

In animals, cholesterol is the key regulator of membrane fluidity. Cholesterol contains a bulky steroid nucleus with a hydroxyl group at one end and a flexible hydrocarbon tail at the other end. Cholesterol inserts into bilayers with its long axis perpendicular to the plane of the membrane. The hydroxyl group of cholesterol forms a hydrogen bond with a carbonyl oxygen atom of a phospholipid head group, whereas the hydrocarbon tail of cholesterol is located in the nonpolar core of the bilayer. The different shape of cholesterol compared with phospholipids disrupts the regular interactions between fatty acyl chains. In addition, cholesterol appears to form specific complexes with some phospholipids. Such complexes may concentrate in specific regions within membranes. One result of these interactions is the *moderation of membrane fluidity*, making membranes less fluid but at the same time less subject to phase transitions.

ALL BIOLOGICAL MEMBRANES ARE ASYMMETRIC

Membranes are structurally and functionally asymmetric. The outer and inner surfaces of *all known biological membranes have different components and different enzymatic activities.* A clear-cut example is the pump that regulates the concentration of Na^+ and K^+ ions in cells. This transport protein is located in the plasma membrane of nearly all cells in higher organisms. The Na^+-K^+ pump is oriented so that it pumps Na^+ out of the cell and K^+ into it. Furthermore, ATP must be on the inside of the cell to drive the pump. Ouabain, a specific inhibitor of the pump, is effective only if it is located outside.

Membrane proteins have a unique orientation because they are synthesized and inserted into the membrane in an asymmetric manner. This absolute asymmetry is preserved because membrane proteins do not rotate from one side of the membrane to the other and because *membranes are always synthesized by the growth of preexisting membranes*. Lipids, too, are asymmetrically distributed as a consequence of their mode of biosynthesis, but this asymmetry is usually not absolute, except for glycolipids. In the red-blood-cell membrane, sphingomyelin and phosphatidyl choline are preferentially located in the outer leaflet of the bilayer, whereas phosphatidyl ethanolamine and phosphatidyl serine are located mainly in the inner leaflet. Large amounts of cholesterol are present in both leaflets.

EUKARYOTIC CELLS CONTAIN COMPARTMENTS BOUNDED BY INTERNAL MEMBRANES

Thus far we have considered only the plasma membrane of cells. Many bacteria such as *E. coli* have two membranes separated by a cell wall (made of proteins, peptides, and carbohydrates) lying in between. The inner membrane acts as the permeability barrier, and the outer membrane and the cell wall provide additional protection. The outer membrane is quite permeable to small molecules owing to the presence of porins. The region between the two membranes containing the cell wall is called the *periplasm*. Other bacteria and archaea have only a single membrane surrounded by a cell wall.

Eukaryotic cells, with the exception of plant cells, do not have cell walls, and their cell membranes consist of a single lipid bilayer. In plant cells, the cell wall is on the outside of the plasma membrane. Eukaryotic cells are distinguished by the use of membranes inside the cell to form internal compartments. For example, peroxisomes, organelles that play a major role in the oxidation of fatty acids for energy conversion, are defined by a single membrane. Mitochondria, the organelles in which ATP is synthesized, are surrounded by two membranes. Much like the case for a bacterium, the outer membrane is quite permeable to small molecules, whereas the inner membrane is not. Indeed, considerable evidence now indicates that mitochondria evolved from bacteria by *endosymbiosis*. A double membrane also surrounds the nucleus. However, the *nuclear envelope* is not continuous but, instead, consists of a set of closed membranes that come together at structures called *nuclear pores*. These pores regulate transport into and out of the nucleus. The nuclear membranes are linked to another membrane-defined structure, the *endoplasmic reticulum,* which plays a host of cellular roles, including drug detoxification and the modification of proteins for secretion. Thus, a eukaryotic cell comprises interacting compartments, and transport into and out of these compartments is essential to many biochemical processes.

PROTEINS ARE TARGETED TO SPECIFIC COMPARTMENTS BY SIGNAL SEQUENCES

The compartmentalization of eukaryotic cells makes possible many processes that must be separated from the remainder of the cellular environment to function properly. Specific proteins are found in peroxisomes, others in mitochondria, and still others in the nucleus. How do proteins end up in the proper compartment? Even for bacteria, some targeting of proteins is required: some proteins are secreted from the cell, whereas others remain in the cytosol.

Proteins include specific sequences that serve as address labels to direct the molecules to the proper location. For example, most peroxisomal proteins end with a sequence, Ser-Lys-Leu-COO^-, that acts as an autonomous targeting signal. The removal of this sequence from a protein that normally resides in peroxisomes blocks its import into that organelle, whereas the addition of this sequence to a protein that normally resides in the cytosol can direct that protein to peroxisomes. A protein destined to pass through both mitochondrial membranes usually has a targeting sequence at its amino terminus. Unlike the peroxisomal targeting sequence, these amino-terminal sequences are highly variable;

no clear consensus exists. They are typically from 15 to 35 residues long and rich in positively charged residues and in serines and threonines. Proteins destined for the nucleus have internal targeting sequences. A typical nuclear localization signal contains five consecutive positively charged residues such as Lys-Lys-Lys-Arg-Lys. The addition of such a sequence to a protein not found in the nucleus can direct it to the nucleus. Other sequences can direct proteins out of the nucleus.

Targeting sequences act by binding to specific proteins associated with each organelle. The determination of the structure of a protein, *α-karyopherin*, that binds to the nuclear localization signal reveals how the protein recognizes such a targeting sequence. A peptide containing the appropriate sequence binds to a specific site on the protein. The target peptide is held in an extended conformation through interactions between the target peptide backbone and asparagine side chains of the α-karyopherin while each of the basic residues lies in a deep pocket near the bottom, lined with negatively charged residues. Proteins that bind to the other targeting signal sequences presumably also have structures that allow recognition of the required features. Note that we have considered only how proteins are marked for different compartments. Later, we will consider the mechanisms by which proteins actually cross membranes.

MEMBRANE BUDDING AND FUSION UNDERLIE SEVERAL IMPORTANT BIOLOGICAL PROCESSES

Membranes must be able to separate or join together to take up, transport, and release molecules. Many take up molecules through the process of *receptor- mediated endocytosis*. Here, a protein or larger complex initially binds to a receptor on the cell surface. After the protein is bound, specialized proteins act to cause the membrane in the vicinity of the bound protein to invaginate. The invaginated membrane eventually breaks off and fuses to form a *vesicle*.

Receptor-mediated endocytosis plays a key role in cholesterol metabolism. Some cholesterol in the blood is in the form of a lipid-protein complex called *low-density lipoprotein*

(LDL). Low density lipoprotein binds to an LDL receptor, an integral membrane protein. The segment of the plasma membrane containing the LDL-LDL receptor complex then invaginates and buds off from the membrane. The LDL separates from the receptor, which is recycled back to the membrane in a separate vesicle. The vesicle containing the LDL fuses with a *lysosome*, an organelle containing an array of digestive enzymes. The cholesterol is released into the cell for storage or use in membrane biosynthesis, and the remaining protein components are degraded. Various hormones, transport proteins, and antibodies employ receptormediated endocytosis to gain entry into a cell. A less advantageous consequence is that this pathway is available to viruses and toxins as a means of entry into cells. The reverse process—the fusion of a vesicle to a membrane—is a key step in the release of neurotransmitters from a neuron into the synaptic cleft. Although the processes of budding and fusion appear deceptively simple, the structures of the intermediates in the budding and fusing processes and the detailed mechanisms remain active areas of investigation.